Vanessa
Bares All

www.penguin.co.uk

Vanessa Feltz

Vanessa
Bares All
Frank, Funny and Fearless

bantam

TRANSWORLD PUBLISHERS
Penguin Random House, One Embassy Gardens,
8 Viaduct Gardens, London SW11 7BW
www.penguin.co.uk

Transworld is part of the Penguin Random House group of companies
whose addresses can be found at global.penguinrandomhouse.com

First published in Great Britain in 2024 by Bantam
an imprint of Transworld Publishers

Copyright © Vanessa Feltz 2024

Vanessa Feltz has asserted her right under the Copyright,
Designs and Patents Act 1988 to be identified as the author of this work.

Every effort has been made to obtain the necessary permissions
with reference to copyright material. We apologize for any omissions
in this respect and will be pleased to make the appropriate
acknowledgements in any future edition.

As of the time of initial publication, the URLs displayed in this book link
or refer to existing websites on the internet. Transworld Publishers is not
responsible for, and should not be deemed to endorse or recommend,
any website other than its own or any content available on the internet
(including without limitation at any website, blog page, information page)
that is not created by Transworld Publishers.

A CIP catalogue record for this book
is available from the British Library.

ISBN 9780857506504

Typeset in 12/15.5pt Minion Pro by Jouve (UK), Milton Keynes
Printed and bound in Great Britain by Clays Ltd, Elcograf S.p.A.

The authorized representative in the EEA is Penguin Random House Ireland,
Morrison Chambers, 32 Nassau Street, Dublin D02 YH68.

No part of this book may be used or reproduced in any manner for the purpose of
training artificial intelligence technologies or systems. In accordance with Article 4(3)
of the DSM Directive 2019/790, Penguin Random House expressly reserves
this work from the text and data mining exception.

Penguin Random House is committed to a sustainable future
for our business, our readers and our planet. This book is made
from Forest Stewardship Council® certified paper.

Zekey, Neroli, AJ and Cecily, I hope you will agree being one of my descendants is a hoot. I insist on deploying you to form a human chain in all weathers. I believe surviving an ad hoc obstacle course constructed by me is the cure for all that ails you. I have taught you rude limericks with which to turn strangers into chums. Never forget what that fellow from Australia did to his buttocks. Rhymes will stand you in superlative stead when conversation fails. I will continue to invite you to formal soirées at which you will be required to perform a party piece and be announced by a liveried flunkey. I will insist you form a collection, so you have something to be excited about at car boot sales. I can't promise not to whoop and cheer at your school concerts. I will sign birthday cards 'Anonymous but v. famous'. If you ever get round to reading this book, you will see I have notched up a few adventures and fitted in big fat dollops of fun and my most treasured moments are always with you four darlings and your beautiful mothers Allegra and Saskia. I love you with all my heart.

Contents

Prologue — 1

1. What did that baby say? — 5
2. I ended up the second-most famous fat person in the UK after Dawn French — 11
3. Matchmaker, matchmaker — 18
4. *Jackie* magazine – practise French kissing on the back of your hand — 23
5. Romeo and Juliet — 29
6. 'For this you went to Cambridge?' — 35
7. 'Darling, you have delighted us long enough' — 42
8. Grandma Sybil sends me to Casualty — 44
9. Quick! Book the caterers! — 48
10. Going to the synagogue and I'm going to get married — 54
11. A little bit of dysentery for you, darling, a little bit for me — 58
12. First comes love, then comes marriage, then comes Vanessa with a baby carriage — 62
13. Fat, fatter, fattest — 66
14. Working nine to five — 69
15. There is no parapet moment — 72

16	Fergie's toe sucking	77
17	*This Morning* with Richard and Judy	80
18	Whatever you do, don't admit you're enjoying yourself	83
19	Check your husband's testicles for lumps. Here's Vanessa Feltz	85
20	Have you ever considered stand-up?	88
21	Opening night	92
22	Every Jew is a critic	95
23	Leonardo DiCaprio, Omar Sharif and Johnny Depp	99
24	New York calling. *Maury Povich* wants you	102
25	*What Are These Strawberries Doing on My Nipples? I Need Them for the Fruit Salad!*	104
26	We want to make you the British Oprah	106
27	Watch Sally Jessy Raphael. Do what she does	109
28	Why make one pilot when you can make twelve?	112
29	Brits can't make talk shows	114
30	All systems go!	117
31	Critics are disgusted	120
32	Overexposure, so soon!	123
33	They want your autograph? Don't be ridiculous	126
34	Welcome to the dead mothers society	128
35	She's fucking huge. What can we dress her in so she doesn't frighten the horses?	131
36	*The Big Breakfast*	134
37	Go to work on a bed	136
38	Damn! I should have shagged Wesley Snipes	139
39	The worst bedfellows – Madonna and Miss Piggy	145

40	The worst of all: Rolf Harris	150
41	Coming up, two great Birds-Eyedeas	155
42	What it's really like to be famous	158
43	Fun facts about being famous	162
44	Co-host *The Jerry Springer Show* in Chicago	169
45	LA calling	173
46	Famous people get to do weird stuff	176
47	Can't we just enjoy it for a moment?	183
48	Life in a rock-star mansion	188
49	Hello, Vanessa, this is the BBC	192
50	The great fall	196
51	That was nothing	200
52	That was nothing	203
53	I do not rule out the possibility of a divorce	205
54	I've never felt so cold	211
55	Piers Morgan, thank you	214
56	Divorce equals demotion	217
57	A quick cautionary tale	223
58	On the bright side, you're gaunt	228
59	*Celebrity Big Brother*	232
60	Oh yes I did!	240
61	Saskia's batmitzvah	246
62	Back at the BBC	250
63	One Hit Wonder	263
64	Gongs, winning and losing, seeking a Vulcanologist	274
65	Gastric band	277
66	Radio 2, *Early Breakfast*	281
67	Darcey Bussell, 'The cha-cha-cha is not your dance'	284

68	All sorts, gastric bypass, Covid, take your pick	289
69	A line of bosses. Has someone died?	300
70	How can I be sixty?	306
71	A tip-off to my daughter	311
72	Pastures new	317
73	I've been out every night for 530 nights	322

Epilogue	328
Acknowledgements	335
Index	337

From: ITV's This Morning

Subject: VANESSA

Yes, can we book Vanessa for International Women's Day?
 Also is she free on Monday?
 We'd like to do an item on RUMPOLOGY and need three volunteers to have their bums read. They would need to show their bum cheeks but could wear a thong.

Prologue

IF ONLY THIS WERE ONE of those books in which the beloved National Treasure looks back contentedly at a curated and constructive life exquisitely lived. You know the type. She had a vision in kindergarten and never once wavered, despite a gruelling incarceration in an exclusive Cayman Islands rehab with Robert Downey Jr, an ill-fated but fascinating fling with David Walliams and a platinum-haired love child with Boris Johnson. Even in her darkest hour she never sprouted a ripple of cellulite, or missed Christmas in Aruba with the Beckhams and, even when the chips – which she never eats – were down, managed to make millions selling vegan tampons and menopause welcome mats hand-woven in Siberia. Her story climaxes in a damehood, personally bestowed at his own request by His Majesty, a bikini cover shoot for *OK!* and an edifying list of illuminating life lessons.

The message sparkles like Swarovski. I am a winner. I am thinner. I have never eaten a chicken dinner. I am rich and multiply orgasmic. Meghan Markle sent me pot number three of her 'limited edition' jam. My line of candles, 'Take a whiff of my butthole', replicating the fresh freesia scent of my bum has already sold out. Make all the right decisions, as I did, and you, too, will be besties with Kim Kardashian and a Booker Prize judge.

I wish I could tell you with hand on heart and a straight face how I turned tragedy into triumph and found true love, my

G-spot and a fail-safe alfalfa recipe. Alas, that's not my story. Over the years – you can check ancient copies of *Heat* – I've been handed a brace of bruising life lessons but, in the way most of us do if we're honest enough to admit it, I've mostly managed to ignore every one of them. It's tough to tinker with your programming. If you're brought up to think a slice of apple strudel is the right and proper cure for a grazed knee it takes more than willpower and therapy to stop yourself mainlining chocolate mousse when your mother dies.

Don't think I haven't listened avidly to all manner of excellent advice. I have, especially my own. I've been an agony aunt for the past thirty-two years everywhere from *Woman's Own* to *This Morning* to the BBC. Does your loved one suffer from a premature ejaculation problem so acute you only have time to lie back and think of Eng? Are you worried about your asymmetrical labia or your son's Kylie tattoo? Do you fancy your mother-in-law? Your next-door neighbour steals your knickers from the line and you rather like it? Don't worry, I'll gallop to the rescue. Unfortunately, I just haven't always seemed able to rescue myself.

I was brought up to believe in a fairy tale. If you were a good girl, if you were clever, hard-working, clean and nice, polite and punctual, washed behind your ears, dried between your toes and moisturized ferociously morning and evening, you would live happily ever after. If you were a devoted diligent wife – suspenders and stockings in the boudoir, pinny in the kitchen, chicken soup bubbling on the hob, children jabbed, filled with fluoride drops and fluently chanting their eight times table – and if you earned a mountain of dosh to keep the family financially afloat but never mentioned it so your husband wouldn't feel emasculated, he would adore you for ever. Never be drunk, late, lazy or rude. Never be ungrateful or unfaithful. Swallow as if you're enjoying it, and one day you'll be seventy-five, Gok Wan will dress you in flattering gold lamé for your granddaughter's wedding and you'll sail to the top table in the marquee on the arm of your steadfast, loyal, still ravishingly sexy at seventy-seven, solvent, respectable husband.

Gentle Reader, in January 1999 I had a husband I adored, who I thought adored me, two exquisite daughters aged ten and thirteen, not the acquired-taste kind, the gorgeous variety you'd hire from Central Casting if they weren't your flesh and blood, a rock-star mansion with an indoor pool and eleven marble bathrooms – you try giving them a quick once-over with Cif – a stonking golden-handcuffs contract with the BBC for a show with my name on it, and a glittering Yellow Brick Road to happy-ever-after all mapped out. By February 1999 my TV show was engulfed in scandal. By August it was over and so (it seemed) was my career. By September the husband I worshipped walked out the door saying he'd keep the air miles and, without a word of explanation, ended our marriage. To nick a catchphrase from the Tories, I tried to 'Build Back Better'. Who'd have thought, sixteen years into a relationship with a boy-bander ten years my junior, I'd be single and starting over at sixty-two? Don't answer that. Everyone – including you, Gentle Reader – saw the writing on the wall except me.

You think you're in the driving seat of a surrey with or without a fringe on top, heading off for a supercalifragilisticexpialidocious morning with the corn at elephant's-eye height then, just as you're merrily obsessing about the velvet trim on the living-room curtains, you hurtle headlong into the brick wall of reality. I'm supposed to be reasonably bright. I know chunks of *Hamlet* by heart. How could I sincerely have expected to waltz merrily from success to triumph without some gruesome gremlin of fate turfing me out of the life I'd ordered from the menu? Honestly – and you'll find I have been honest: I can't see the point of writing all this stuff down otherwise – I truly don't know.

Here is one lesson I have learned. You can be funny and faithful, bestow the best blow jobs in the Borough of Barnet, dish up ambrosial spag bol and tell rib-tickling stories about cavorting on the *Big Breakfast* bed with Jim Carrey, but if a man fancies a foray with a cocktail waitress/your best friend/your daughter's piano teacher, you might as well make him a farewell salt-beef sandwich on rye, cut your losses, put your best Jimmy Choos forward and move onwards and upwards. Failing that, glue on your lashes, drum up your gal pals, hit the town running, chain-imbibe piña coladas while dancing round your handbag to 'I Will Survive' and pretend.

1

What did that baby say?

'WHAT DID THAT BABY SAY?' a man asked my mother. He was spooked.

I was nine months old. I had been talking about him. I said, 'I'm nervous of that man.' It was in a shop. I was in my Silver Cross pram. He was elderly, with rosy cheeks, and he'd peered in at me. I didn't like being stared at. I said clearly to my mother, 'I'm nervous of that man.' Today he'd probably be arrested.

Then a mild hue and cry broke out: 'What did that baby say?'

Turn to the last page of this book and you'll find me carrying my youngest granddaughter, Cecily Violet, towards Marble Arch sixty-two years later. Cecily is twenty months old. We are going to an art exhibition. Cecily gives her views on modern art. A woman in the street looks as if she's seen a ghost. She asks me: 'What did that baby say?'

I say: 'She said, "Picasso. My favourite."' The woman looks spooked. I am my mother. Cecily is me. History repeats itself.

When I tell people I detested being a baby they don't believe me, but it's true. Having a baby is wonderful, but being a baby is boring. When you're a baby there's nothing to do but wait. And baby talk is so . . . babyish. You never have a good conversation with adults. They won't gossip. It's all moo-moo and quack-quack.

Did I care that Old MacDonald had a farm or which piggy went to market, which piggy stayed at home, or how much they forked out for that doggie in the window? Trapped in my pram-prison in the freezing cold – in the sixties they stuck babies outside in the 'fresh air' even when a fine coating of snow fell on the coverlet – I plotted escape.

I despised babyhood. When Grandma Sybil changed my nappy, I did my utmost to engage her in diverting pleasantries. 'Have you got a bad back, Grandma? Do you want a cup of tea, Grandma?'

She refused to engage. 'That child is too young to talk. I'm not pandering to it. This is absurd.' I soon discovered there was something worse coming my way: childhood. I wanted to skip the whole baby–child–adolescent routine and go straight to the interesting segment: being a married woman, with pointed ruby-red nails and a Triumph Herald, living somewhere exciting like Docklands.

Until then I had to endure a life of suburban tedium in Totteridge, the Beverly Hills of North London, the swimming-pool and chopped-liver belt. In Totteridge, fathers polished their company XJ6s on Sunday mornings after taking the children to Hebrew classes, and mothers struggled to stay slim at keep-fit classes while making high-cholesterol *lokshen* pudding and apple strudel with a million mouthwatering calories per teaspoon.

In 1962 Totteridge was still in Hertfordshire. It didn't morph into London N20 for years. With the Terenure Country Club, verdant Totteridge Common and roads named after British press barons, Northcliffe and Harmsworth, it was an emerald enclave of gentiles plus a trickle of Jewish families spilling northwards from Hendon and Golders Green. Rivulets of anti-Semitism dribbled just behind the suburban smiles. How did we know we weren't altogether welcome? What about the local South Herts Golf Club's policy? NO JEWS ALLOWED? When I say local, how's this for proximity? Our garden backed on to the course. There was a gate, just behind our greenhouse, which opened on to the eighth hole. My father played golf 'religiously' every Saturday – just not there. He had to drive for twenty minutes to Potters Bar Jewish golf club to tee-off.

Jews were barred from membership at non-Jewish golf clubs in sixties Britain. It was a well-known fact. So well known, there were plenty of Jewish jokes about it. Hymie Goldberg wants to join the local gentile club. He changes his name to Harold Green, puts on a tweed suit and goes to meet the chairman. Name? Harold Green. Age? Forty-two. Religion? 'I am of the goyish faith.' The joke? Hymie's using a Yiddish word 'goyish', which means non-Jewish, to demonstrate that he isn't Jewish. Hymie gets booted out. Here's another: Laybish Cohen wants to join the local gentile club. He changes his name to Leonard Conrad and goes to meet the chairman. Name? Leonard Conrad. Age? Forty-two. Suddenly a heavy paperweight falls off the chairman's desk and on to Laybish's foot. Laybish yells: '*Oy vey!* Whatever that may mean!' The joke? In pain, he shouts a Yiddish exclamation of agony, *Oy vey!* Woe is me!, realizes what he has said and pretends not to understand the phrase he himself has just shouted. Naturally, Laybish gets bundled out of the club.

Frankie 'Give Me The Moonlight' Vaughan – I loved him as a little girl and will never stop loving him though he died in 1999 – was really called Frank Fruim Abelson. He changed his name because sounding less Jewish helped careers. (Both my uncles, one a surgeon, one a lawyer, changed theirs from Ohrenstein to Orton.) Frankie called himself Vaughan because his grandma, accent from *der Haim* (back home, the Old Country), said he was her number 'vorn' (one) boy. He starred in Hollywood movie *Let's Make Love* opposite Marilyn Monroe but didn't respond to Marilyn's invitation to run through the script at her place one night because he loved his wife, Stella. Frankie's house backed on to the opposite side of South Herts Golf Club from ours. He also played golf, but even the name change, Royal Variety performances and leading the crowd, with Rabbi Baum's permission, in Anglican hymn 'Abide With Me' at the FA Cup final, didn't gain him membership.

Meet the Feltz family

At Four Winds, Pine Grove, Totteridge, Herts were my father Norman the 'Knicker King', in the ladies' lingerie business, my mother Valerie, me and my younger sister. (She's a private person so you won't hear any more about her.) My mother was a housewife – though she loathed the term, saying furiously, 'I am not married to the house.' We called my parents' pals 'uncle' and 'auntie'. Uncle Ronald had chemist shops. Uncle Leslie was in furniture, and our neighbour's personalized number plate was 1WMT – number one in the Wholesale Meat Trade. The women – apart from Helen Bloom, who manufactured bespoke lampshades – didn't work outside the home. They'd married in 1959, just missing liberation, emancipation, free love and 'turn on, tune in, drop out'. Hostess trolleys were wheeled from kitchen to dining room to keep the meatballs at a toothsome temperature. My mother kept a leather-bound 'guest' book faithfully recording which cordon bleu dishes she'd served to which group of friends at which dinner party. They had elevenses with their cleaning ladies – my mother and her 'daily' called each other Mrs Feltz and Mrs Parker. They did their best to lunch without eating.

The Yiddish chorus

Nothing I did at Four Winds went without a cascade of criticism. I had my own personal Yiddish chorus – the Jewish equivalent of the Greek chorus providing constant commentary on existential subjects Life and Death. My chorus stuck to the smaller stuff but treated it like life and death. Everything was chorus fodder: what I wore – dressing gown, slippers, what I should have worn – vest, scarf, woollen hat, what I read or thought, who I called on the phone (never, ever before six p.m. when calls were more expensive and never, ever boys: wait for them to ring you!). The chorus insisted I learn to play tennis or bridge, so I wouldn't bore the boys silly. Aloud and loudly the chorus identified my deficiencies. What was that ridiculous expression on my face? Hurry, use concealer on that spot on your chin. Is that lipstick? No? Why not? Not that colour. It's far too old for you. Petal pink. Always petal pink. You don't want to give the wrong impression. But you don't, Heaven forbid, want to give no impression.

This is what I mean

You might think the ordinary, thrice-weekly, act of hair washing would not be worthy of comment, but you'd be wrong. I only had to mention that I was going to wash my hair and the chorus would spring into action.

'What? You're washing what?' says my father, who's reading *The Times*. 'You're washing your hair? Why are you washing your hair now? We're just about to have dinner,' or we've just had dinner, or we're going to have dinner, we're thinking about dinner 'and you're washing your hair now? Why are you washing your hair now? It's Wednesday, you're going to a party on Saturday, so why are you washing your hair now? Valerie, why is she washing her hair now?'

Cue my mother: 'Vanessa, why are you washing your hair?' She could go either way. It might be 'Vanessa, under no circumstances whatsoever should you even think about washing your hair now.' Or possibly 'Norman, focus on Bernard Levin. Why shouldn't she be washing her hair now? This is a good time for her to wash her hair. She's got to do her Latin homework and she hasn't done her piano practice or written a thank-you letter to Auntie Hilda.'

As if by magic Grandma Sybil rings: '*Nu?* [What's going on?] Vanessa? What's happening? She's washing her hair now? Hasn't she just had a cold? She's had a cold, and she's risking pneumonia washing her hair? Valerie, why is she washing her hair now? Tell her I said absolutely not.'

Five minutes later, Grandma Babs is on the line: 'Vanessa is washing her hair? If she wants to wash her hair, let her wash her hair. She should wash it. If she feels comfortable washing it, happy to wash it, she should wash it. I think she should wash it. And, *k'neyna hara* [may the Evil Eye not punish her], she has such gorgeous hair. It comes from my mother Malka.'

Some of my chums had a complimentary Yiddish chorus pouring out streams of sweet nothings. 'Darling, you look beautiful. What a picture! That dress is adorable on you. Wash your hair? No need. Let me pay for you to go to the hairdresser. Your hair is lovely. You're lovely.'

I had occasional outbursts of approbation, but my chorus was a self-esteem shredder. Yet would you believe I miss it terribly? For all its cacophonous craziness and casual cruelty, I was never, ever, even when I wanted to be, unseen or unheard. The constant commentary enforced the belief that I was visible, alive and a host of people cared passionately about whatever I was doing.

Even now in 2024, when I choose hand towels for the guest loo and there's no one to pipe up, 'Why the pink? What's wrong with the turquoise?' I feel a devastating loneliness. I'm choosing towels by myself. I can buy whichever I fancy. There's no one telling me I'm wrong. The silence is crushing. I can't bear it. I desperately miss my Yiddish chorus. What is the point of doing anything without their judging panel? Now I can select the towels I want without their focus group, I don't have the heart to buy any at all.

2

I ended up the second-most famous fat person in the UK after Dawn French

Here's how it happened

I WASN'T ONE OF THOSE chubby children finishing off the birthday cake and looking for another biscuit, the cherubic pint-sized putti, who make you think, Aah! Little foodies in the making. Aren't they adorable?

You may not believe this, but I was a skinny little girl, not at all interested in food. I wouldn't eat cheese or sponge cake. My mother poured tinned pears over lamb chops to 'trick' me into chewing them. I was a finicky, picky, reluctant eater. At primary school, the cooks knew I always asked for an extra-small portion. Food wasn't my thing. I didn't daydream about dinner, or lunch, or chocolate buttons. I hardly thought about eating at all except how to avoid clearing my plate.

I was slim till – vengefully early at the age of eight – puberty struck. I started what they used euphemistically to call 'developing'. My mother was horrified. I was whisked straight to London's premier paediatrician, Dr Lionel Balfour-Lynn. Barely concealing what sounded to eight-year-old me like a toxic mixture of panic and revulsion, she threw herself on the mercy of the eminent doctor. 'Doctor! Help me! It's Vanessa! There are swellings. In the "chestal" area. Is it? Could it be [loud stage whisper] cancer?'

Lionel Balfour-Lynn laughed, in a London's-premier-paediatricianly way: 'Ha! Ha! Ha! Cancer! Far from it. Vanessa is in rude health. She's developing. These small mounds, Mrs Feltz, are the beginnings of her breasts.'

Breasts! At eight! The Yiddish chorus cranked up several gears. Grandma Sybil: 'Bosoms! The child is eight. Don't be ridiculous, Valerie. Get a second opinion.' (We all know what happens if you get a second opinion, though. Doctor: 'Mrs Cohen, you are pregnant.' Mrs Cohen: 'Pregnant? I want a second opinion.' Doctor: 'You are also ugly.')

Grandma Babs: 'It's perfectly normal, *boobelah* [sweetheart]. Cousins Minnie and Bessie wore brassières at eight. We're a family of early developers.'

My father, Norman the Knicker King, sobbing: 'Valerie, Vanessa's our baby. This is too soon. I'm not ready. I'm only thirty-two. I'm much too young to have a daughter with breasts.'

Grandma Babs took me to Pullen's in Temple Fortune. My nascent bosoms were confined in the Littlest Darling bra by Berlei. No one asked me how I felt. If they had, I'd have said I was consumed with guilt for putting them all through the emotional wringer and thoroughly embarrassed to be a year younger than everyone else in my class but the first to wear a bra.

My mother gift-wrapped her own weight problem and handed it to me

I'm not a psychologist or a therapist. I can't read my late mother's mind. I did try to talk to her about this several times, but she vehemently refused to discuss it. So what I'm about to tell you is only my opinion, not fact. I think what I am about to write is true, but I may be wrong. After trying to work this out for decades this is my conclusion. My mother handed me her own weight problem. I don't think it was on purpose. I'm not sure she even did it consciously, but she pretty much put her weight problem in a box, tied it with a big pink ribbon and presented it to me, like a gift. I'll tell you exactly what happened. You can decide if you agree with my analysis.

Between the ages of eight and nine, the shape of my body started to change. My waist went in a bit. My hips widened a little. My bottom and tummy became infinitesimally rounder. I grew taller – not much, I'm still only five feet two and a quarter – my skin was a tad oily, there was a light dusting of spots on my T-zone, and my hair, until then straight and silky, grew greasier and frizzier. When my mother saw my body begin to change, she panicked. I didn't look like Alice in Wonderland any more: I looked like an almost-teenager with the potential to be podgy. She didn't like it and was determined to do whatever it took to bring it to a halt.

My mother had always had a weight problem – or, at least, her own mother, Sybil, thought so. Sybil was five foot seven and skinny. Her wedding photographs confirm she weighed six and a half stone as a bride. She was also ravishing, rather like the actress Shirley MacLaine. I remember as a little girl seeing Grandma Sybil walk into the room and thinking, Oh, my goodness, my grandma is beautiful! My mother was pretty, with curly blonde hair and cornflower-blue eyes, but her shape was different. She'd inherited it from her father, stocky, portly, round-bellied, five-foot-six furrier Poppa Willie. She was shorter than her mother, thicker-set and pear-shaped with a big bottom. Her own grandpa exclaimed: 'Valerie has *k'neyna hara* legs!' (Legs so substantial you'd want to ward off the Evil Eye by diverting attention from them. Not a compliment.)

Grandma Sybil was disappointed with the well-upholstered look of my mother. She'd exclaim: 'Valerie, what are you wearing? You look like a sofa in that. It's not doing you any favours. You look like a settee, darling. Valerie, what are you eating? No apple pie for you tonight, Valerie!'

You'd never have known it to look at her, but my mother was locked in internecine warfare with her waistline. She wasn't a size ten or a twelve, probably more like a generous size fourteen or sometimes a sixteen. Forty-seven per cent of British women are a size sixteen or over. She wasn't plump or borderline obese. She was just chubbier than she'd have liked to be, chubbier than her mother, chubbier than was chic. Remember, we're talking about the sixties and seventies, when Twiggy, Jean Shrimpton and skinny girls with flat chests ruled supreme. My mother's Totteridge friends, Pat, Loretta, Margaret and Gillian, somehow managed to be svelte with boyish snaky hips and thighs that didn't meet in the middle. Not my mum. She wasn't built that way and once, in her only honest moment of revelation, admitted to 'eating an entire dinner while I'm cooking your dinner' and keeping secret packets of Revels in the car's glove compartment. When my body began to change, I think my mother thought: Oh, my God! Vanessa is going to get fat!

I'll stop speculating about my mother's thoughts/motives/inner demons now. Instead, I'll simply tell you what she did. I was nine. I fitted comfortably into clothes for nine-year-olds. Nevertheless, she began to curb, quite drastically and publicly, what I was allowed to eat. We'd be having dinner, and my mother would announce to the rest of the family, 'We're all having the soup, with *kreplach*, *kneidlach* and *lokshen* [wontons, dumplings and vermicelli]. For you, Vanessa, half a grapefruit.' She'd declare: 'It's chopped liver for us tonight. Vanessa is having a slice of melon.' I was nine. It felt like a punishment, and I didn't know what I'd done to deserve it. I hadn't been caught shoving a whole box of contraband chocolates into my mouth. It made no sense. One day I wasn't thinking about food at all and the next, without explanation, I was put on some sort of diet. I didn't say anything because I didn't know what to say.

My father, who had a voluminous vocabulary and an insatiable urge to unleash it upon an array of topics high and low brow, was uncharacteristically silent. He didn't ask what was going on or why. He never looked at the diminutive daughter sitting next to him and came to my defence. I don't know why he was quiet. Maybe my food intake didn't directly concern him, and he was too swept up in his all-consuming narcissism to notice. Maybe he saw what was going on but sensed the weight issue had deeply visceral permutations for my mother and couldn't face raking up controversy. I do know he could and should have said: 'Valerie. She's only nine years old! What are you talking about? Why can't she have some soup?'

The regime was sudden and shocking. It left me with a new sensation: hunger. I'd never been hungry before. I'd never thought: Everyone's asleep. Is there anything in the fridge I can sneak without being found out? I'd never had the desire to pick the dried banana out of the muesli and crunch it. I'd never begged my sibling to ask for an orange between meals so I could eat it. Now I did all those things. And I was still hungry all the time.

At ten I went to senior school and took a packed lunch made by my mother. It consisted of two slices of white bread and between them something so minimal it was almost transparent – one wafer thin slice of processed meat or a light hint of Marmite or Dairylea – and an apple. I didn't have time to eat breakfast in the mad dash to catch the High Barnet coach to Haberdashers' Girls' School, so by first break at eleven o'clock, I was ravenous. I'd demolish my lunch in seconds. The other Jewish girls' packed lunches were veritable banquets. Anne-Marie Gold's mother provided two different flasks for beverages hot and cold. My pals produced pretzels, slices of strudel, extra capacious lunchboxes overflowing with slices of *wurscht* (kosher salami), legs of chicken, lamb chops, bagels lavishly layered with smoked salmon. So abundant were their feasts, they furnished an ample elevenses followed by a substantial lunch at one. I didn't have any cash. My dad didn't believe in pocket money, so I couldn't go to the tuck shop and supplement the insufficiency with a Mars bar or a Twix.

By the time I got home at five, I was ravenous. I was growing. I hadn't eaten since eleven o'clock. My mother was adamant. No food would be served. If she was worried about calories, she could have allowed me a salad or Ryvita, or baked beans on toast, but she didn't. Her words still send a sad shiver down my spine. 'Dinner is at seven. You'll just have to wait. If you're really "starving" eat an orange.' Pine Grove was at the top of a steep hill. There was no shop to nip to. I didn't have any money anyway. Like any normal hungry child, I'd have rummaged through the biscuit tin, but my mother put a lock on the biscuit cupboard and hid the key. If I'd unearthed the hiding place, I'd have used it. She was excellent at hiding. I searched and searched for that key but never found it.

My attitude to eating changed. Whenever I managed – at the cinema, a party, on my birthday – to get hold of biscuits or a packet of sweets, I devoured them instantly. I couldn't save them because I didn't know when I was ever going to be given more. I became one of those people who eats a whole tube of Love Hearts in two minutes. I chain-consumed anything sugary the way chain-smokers puff cigarettes, with no time to savour, no chance of enjoyment, just grim relentlessness. There was no chance of my sucking a couple of sweets now and putting the rest in my pocket for later. You need to be relaxed to delay gratification. I was, as the Yiddish saying goes, permanently on *shpilkas* (edgy). Eating was urgent. I was forced to find wily ways around the restrictions. I'd offer to clear the table and while ferrying the roast potatoes from dining room to kitchen, shove as many down my throat as I could, swallowing them without even chewing. It was a way of feeling full. It was also a way of getting back at my mother.

It launched me on a cycle all too familiar to anyone with an eating disorder or addiction of any kind. You eat/drink/gamble/snort/shop to make yourself feel better. Yet as you are doing it, you know, vividly and painfully, you are already making yourself feel a whole lot worse. If food is your crutch and curse, it's unavoidable. Eat and you make yourself fatter. Don't eat and you die. You can't go cold turkey on cold turkey. You must eat, but the confusion caused by your mother frowning every time you lift a fork towards your face can't be filled by trifle or Black Forest gateau. I was hungry and bewildered and never mentioned it to anyone.

The Yiddish chorus stayed silent

Why, you might wonder, did the yapping Yiddish chorus zip it when finally something cried out for commentary? Why didn't the grandmas swoop in and nip this *meshuggas* (madness) of my mother's in the bud? Who knows? Could they have been scared to bring something so unnervingly Freudian to the surface? Could they have agreed keeping me slim was the surest route to the ultimate goal – marriage? I can't explain their averted gazes and passive acceptance. I know they loved me, but they didn't bestir themselves to make sure I was properly fed.

3

Matchmaker, matchmaker

> 'Matchmaker, Matchmaker, make me a match
> Find me a find, catch me a catch'
> *Fiddler on the Roof*,
> lyrics by Sheldon Harnick

IT WAS 1972. I WAS a year ahead at school. Women all over the world were burning their brassières in ceremonial sacrifice. At Four Winds we never alluded to such treachery. Grandpa Mo created the 'Laura' bra, a 1950s bosom-hoicking revelation. I had my eye on a boy in my Hebrew class.

Reverend Plaskow would intone an Old Testament verse in Hebrew. We'd chant back the English translation. There was something about the way Jonathan Malka yelled, 'And her rival vexed her sore', something about Hagar getting on Abraham's Sarah's last nerve, that sent Leapy Lee's little arrows straight to my heart. I had to be in love because I knew exactly what I wanted from life. I wanted to get married. Make babies. Make charming love with and apple strudel for my Byronic beguiling husband and live happily ever after. Come to think of it, that's still what I want – minus the baby-making.

When someone asks me what my home life was like as a child,

I say: 'I grew up in *Fiddler on the Roof*.' I'm not joking, and if you haven't seen it, see it. You're right, my family weren't impoverished nineteenth-century peasants eking out a living in the imaginary *shtetl* village of Anatevka. We didn't burst into song when the horse pulling our milk-cart went lame, we didn't cut loose with bouts of exuberant faux-Russian dancing. What we had in common with Tevye the milkman – played on stage and screen by the irresistible Topol – and his five desperate daughters was a daily, all-consuming drama: the urgent need to marry me off.

In our house marriage wasn't a topic to be deferred till I was old enough to have sex, join the army and vote. Before I was severed from the placenta, my parents had decided on the colour of the bridesmaids' dresses and matching table centres (peach), crossed the Goldfinkles off the guest list and selected the kosher caterers. The heat was on. The phrase 'under pressure' to persuade a suitable suitor to pick me doesn't come close to describing the overpowering reality.

The carrot

Grandma Babs and Poppa Mo

It's easy to see why. I had two glorious grandmas, Sybil (maternal) and Babs (paternal). Grandma Babs met Poppa Mo when she was twelve and lending a hand on her mother's underwear stall in the East End's Petticoat Lane market. He was handsome and ambitious. His father, a tailor, sat on the table with his legs crossed and a tape measure round his neck altering suits for the gentry. Mo wanted more. He 'looked just like Clark Gable only much more dashing', rocked a sultry swagger, fought in the Battle of Cable Street, was destined to become an officer in the British Army and chair a public company. Babs (her real name was

Miriam but she was the baby of the family, so everyone called her Babs) was tiny and sweet with rosy cheeks and a penchant for D. H. Lawrence. Of course, they fell in love.

Grandma Sybil and Poppa Willie

Poppa Willie spotted svelte Grandma Sybil front of house in her father's fur showroom. Someone told him she 'was the smartest girl in the East End'. He asked her father for her hand. She was a bride at nineteen. She said jolly jokester Willie 'brought laughter to their home'. He did. They lived happily ever after.

Valerie and Norman

My father was the first boy to French-kiss my mother. She was seventeen. He dumped her soon after. She was proposed to by scores of besotted suitors. She couldn't say 'yes' to those upstart pretenders. She was holding out for the real thing, Norman, captivated by his theatrical antics with a silver-tipped sword-stick and nonchalant habit of peppering his conversation with flurries of Talleyrand, Yeats and Louis MacNeice. At twenty-one, he nearly died in a car accident. She was reading history at the LSE and had too much self-respect to visit him in hospital, but she wrote him a ladylike letter of concern. He responded. They married when she was twenty-one and he twenty-three.

I was reading *Romeo and Juliet*, *Pride and Prejudice* and watching *Peyton Place*, starring Ryan O'Neal as Rodney Harrington. With all that true love just sitting there around the Friday-night table, how could I not be dreaming of a man with Kafka's haunted, hunted expression and my father's abundant raven hair whisking me up the aisle and into a three-bedroom semi in Hampstead Garden Suburb?

Girls all over the world were realizing they could become hod carriers, firemen (the word 'firefighters' was on the cusp of being coined), astronauts or touring puppeteers. Barbie qualified as a vet. Marriage was strictly for dinosaurs and demoted from female fantasies. Jewish girls, on the other hand, were tucking their A levels and degrees into the dresser drawer, praying Yente – *Fiddler on the Roof*'s Matchmaker – would put her skates on and hurry up with that match.

The stick

Let's think crudités. You've had the carrot. Now for the stick. It wasn't all romance, a honeymoon in Eilat (where the sun takes its winter holiday) and being clutched to the chest of a David Baddiel lookalike. There was a hideous spectre looming on the horizon. So horrifying was this prospect, it was spoken of only in whispers. *What if no one ever chooses me? What if I am never snapped up? What if I am left languishing on the shelf?*

Jews do not believe in Hell – a Miltonic dystopia belching fire and brimstone where you suffer for your sins for all eternity. We have our own torture chamber right here in a synagogue hall near you. It is called: THE SINGLES' TABLE AT BARMITZ-VAHS. That's where the women who've failed to haul home a husband are seated at celebrations. It's where the 'odds and sods' end up, the women no man wants, the female chosen people who have ended up the most pitiful of creatures: the unchosen.

I grew up not with the fear of God but a deadly terror of ending up on the singles' table with the odd sods. As a child I was darkly warned a seat there was the ultimate public humiliation. Now I'm older, wiser and reluctantly single, I still struggle not to feel naked dread at the fate worse than death. If you're marooned on the singles' table for six courses, plus petits fours, everyone, including your exes and women you were at school with, knows nobody wants you. Nobody chose you. You are unclaimed. And that's why you are sitting at the undesirables' table too near the toilets. What's more, no one believes you pitched up on that table because you were so busy doing other stuff – your doctorate in quantum physics or creating the Covid vaccine. It's irrelevant. No one buys the idea you just didn't get round to getting married but it's on your to-do list. The stigma is glaring neon. You're enveloped in the cloak of shame. Everyone can see you're next to the creepy cousin from Glasgow with seven cats and an overdue nose job.

To this day I grit my teeth and square my shoulders if I'm invited to a barmitzvah. The excruciating embarrassment of being assigned to that table – my rightful place as I'm single – still haunts me. It's no exaggeration to tell you fear of the seating plan cemented me to my ex for nearly seventeen years. People say airily: 'Being on your own is great! It's empowering! You can be free and autonomous.' I say you're right. Absolutely. I agree. It's true. I am complete. I don't need a bloke, especially one who buys me a birthday present at the hotel shop that comes up on my bill at the end of the holiday for which, of course, I am paying. I am enough on my own. I get it. I've tried, memorably in early 2023 with three different therapists a week, to shake the 'I cannot exist without a man' mindset. It's just hard to feel that way when the opposite has been pummelled into your brain.

When there's a fox in the garden and there's no one to turn to and say, 'There's a fox in the garden,' I don't like it. It still feels unnatural. The sound of silence is deafening. I'm genetically constructed and emotionally hardwired to sleep next to some soul snoring and not on my own. But I'm trying my hardest to see sense, man up and master that most clichéd schmaltzy and elusive skill – feel free to make a vomiting noise – loving my own company!

4

Jackie *magazine – practise French kissing on the back of your hand*

Pete Tong

PACKAGE HOLIDAYS WERE BASIC IN 1972. It was all 'Cosmos this way!', bossy reps in blazers and being forcibly booked on tours of the artificial-pearl factory. The setting was Cala d'Or in Majorca, the hotel the four-star Toucan. Astoundingly, my parents had taken the reckless step of schlepping off on yet another coach to see the stalagmites (or were they stalactites? I don't care. Don't tell me) in what I knew must be the excruciatingly boring Caves of Drach, leaving me to my own devices at the hotel. This was atypical. I was supervised minutely morning, night and noon. It's hardly surprising the moment I sniffed freedom I went for it. Or should I say 'for him'? He was lurking, or possibly larking about, in the pool. He was tall – well, taller than me – dark and Hollywood handsome, with what I still swear were lake-sized violet eyes. His name was Peter. He was twelve with an irresistible smile. Frankly, I didn't try resisting. We hit it off instantly. We splashed and chatted. Flirtation ensued. We kissed – teeth clamped shut, no tickle of tongues, his soft lips pressed against mine.

On my parents' hot, irritable return from the pointless gawp at the tites or mites, news of the lip-locking was tittle-tattling round the sunbeds. They were livid. Words like 'reputation', 'can't be trusted' and 'Off to your room, young lady' bounced off the buffet. Peter and I were inseparable until the plane landed at Luton and we were prised apart by our cruel parents. He visited once. He sent me letters and a ring. His beautiful mum had pictures of us together. Years later, when everything really had gone resoundingly Pete Tong, I saw a picture of the now world-famous DJ whose name had become synonymous with the phrase 'tits up'. It was He – my Peter Tong, the package-holiday prince who charmingly bestowed my first ever kiss and, though I never kiss and tell, I'm here to tell you there was absolutely nothing Pete Tong about it.

A Sponsored Silence

Yom Kippur. The Day of Atonement. The most awe-laden of the Days of Awe. The superlatively solemn Jewish festival in a – let's face it – pretty humourless calendar. The one upon which we pray to be inscribed in the book of life for another year and, to push the Lord into including us, survive twenty-five hellish hours without food or water while symbolically beating our breasts to show penitence for twelve months of sins.

Obviously, when you're eleven the whole thing is a hoot. First, you don't have to fast till you're twelve (for a girl, thirteen for a boy, not fair but, as my father said every hour on the hour, 'Vanessa, life is not fair!') so you're not hungry and headachy, like your grouchy parents. You get the day off school. You get to gaze down at boys from the ladies' gallery in the synagogue and imagine being married to the one with the fewest pustules. You get to read the list of sins and, in my case, wish you'd committed at least some of the most intriguing. You get to read the line about how 'lying with a man as if with a woman is an abomination' and laugh dementedly till your mother swats you with her prayer book.

And in 1973, I was looking at mouth-watering pictures of a groaning smoked-salmon buffet smuggled into synagogue by Sharon Anisfeld, whose dad Marcel was the purveyor of smoked salmon to the cognoscenti and knew a groaning buffet when she saw one. Parched mouths were watering at the gefilte fishball tower, the smoked-herring luge and the life-sized sculpture of the barmitzvah boy in egg and onion. Photos of food. In *shul* (synagogue). On Yom Kippur. We could hardly bear the wickedness.

On Yom Kippur there's no earthly frippery allowed. We don't drive. We don't work. We don't watch TV or listen to the radio. We don't wash, except face and hands. We don't wear leather shoes. So how could the news of a surprise attack on Israeli territory by Arab coalition forces possibly have permeated the sacred citadel of Woodside Park Synagogue? Somewhere in the sanctum a secret TV viewing renegade lurked. The congregation quivered. No one was supposed to know, yet no one could pray for salvation in the customary hungry, headachy, dry-mouthed, bored-stiff yet sporadically awe-stricken manner. How serious was the invasion? How many lives lost? Where could the adults give blood/money/plant even more trees in Israel? They were meant to be fully focused on pleading for their own lives and at least a couple more months for Auntie Sadie, stricken with cancer, poor woman. Instead, they were worried sick about their nephews and brothers-in-law, cousins and uncles. It was 1973, for God's sake. The Holocaust had only been over for twenty-eight years. Worried sick doesn't even begin to do it justice. This was a pestilence of angst. Except, of course, if you were only eleven and the Yom Kippur War meant unprecedented hitherto prohibited access to boys and such sumptuous social magnificence as a Sponsored Silence chez the aforementioned Anisfelds.

The grown-ups were consumed with the Middle East crisis. My father collided with Frohwein the kosher butcher's van in Willesden High Road so vividly was he imagining himself heroically driving a tank in the Golan Heights. Obviously we children couldn't just watch *Blue Peter* and ignore the tumult. We were forces to be mustered, and Marcel and Irene Anisfeld were the first to muster us. We were given forms, told to secure pledges from family and friends and summoned to attend the mysteriously entitled 'SS'.

I had no problem persuading nearest and dearest to cough up. They knew I hadn't stopped talking since birth. Laughing at the absurdity of suggesting I might put a sizeable sock in it, Uncle Clive merrily signed over ten pounds for every hour I managed to stay shtum. Ten quid was a fortune in 1973. You could buy a detached in The Bishops Avenue for £70K. Uncle Clive, renowned maxillo-facial surgeon and bestower of nifty nose jobs on most of North-west London, had no qualms. He reckoned I'd notch up five minutes of quietude max. Chatterbox Feltz at a Sponsored Silence – a risk-free promise. My form filled impressively. The stakes were high. The pressure was immense. Israel's future was riding on it. Utter one syllable and I'd betray my already betrayed people. I was eleven, for God's sake, just a few months away from Bat Chayil year when, Jewishly, I'd become a fully fledged woman. If ever the time had come to shut the beep up, this was it.

I nervously rang the Anisfelds' bell and crashed into an impenetrable wall of deafening silence. The quiet was mind-blowing. The smoked salmon bagels – we pronounced it beigels, to rhyme with 'eyefuls', because we came from Plotsk but I want you to know what I'm talking about – were ultra de luxe. My father said Marcel rose at dawn to catch the wild salmon personally. Stifling the urge to exclaim something, anything, I took in the vista. Boys were everywhere, breathing, stuffing in gefilte fishballs, pretending to do their homework, combing their quiffs – we worshipped the fifties in the seventies, I can't remember why – and staring hungrily at . . . me! Thank God! It was happening. It had taken a hideous conflagration in the Promised Land, but it was finally happening.

Guess what? It's easy to stay silent when you're snogging. A mouthful of mouthful means you can't say a word and there's nothing at all you can think of saying anyway. When you're eleven and lip-locked with Paul Nathan, the boy you've adored from afar – well, eight doors away – since you were pushed around Totteridge Green in adjacent prams, French-kissing isn't just whirring tongues. It's brain-freeze, metaphysical, spiritual, stream-of-consciousness, best-thing-that-ever-happened-to-you magic. I loved it . . . silently. We did lots more of it. For hours. And hours.

Uncle Clive had to pay me seventy pounds in hard cash. My own parents coughed up thirty-five. Neighbours winced as I showed them my signed form confirming I hadn't uttered a peep for seven hours solid. I earned a thwacking hundred and eighty-five pounds. Like Sarah, Rebecca, Ruth and Queen Esther, I'd done my people proud.

Weeks later I received a handwritten letter from the actual Chief Rabbi. No one writes to an eleven-year-old about anything. I was ecstatic. I read and re-read it and kept it in my dressing-table drawer for years. It said in scratchy Quink ink: '*Mazeltov!* A sponsored silence and such results!!'

Vanessa the Undresser

I suppose this is the time to tell you I was known at school as 'Vanessa the Undresser'. I wasn't and I didn't but it rhymed and I had a wiggle in my walk, black tights, forbidden high heels I hid underneath a rhododendron bush, a red scarf hand-knitted by my mother long enough to be wrapped round my neck and trail on the ground either side, which I twirled as coquettishly as a flamenco dancer with a fan.

An enterprising fifth-former came to trim Grandma Babs's privet. 'You might know my granddaughter Vanessa Feltz?' The fellow blushed scarlet to the tips of his earlobes, spluttered, 'Vanessa the Undresser!' and disappeared into the greenhouse. When the Yiddish chorus raised the subject, I feigned ignorance.

5

Romeo and Juliet

Prince Charmless

ON MY THIRTEENTH BIRTHDAY, 21 February 1975, I fell truly, madly, deeply in love. His name was Prince Charmless. I noticed him on the High Barnet coach to Haberdashers'. He was fifteen and a half and carried a copy of T. S. Eliot's *Four Quartets* in his jacket pocket. He was brooding and moody, with Heathcliff hair, furrowed brow and a supercilious air just like my father's. We broke up for half-term. School finished at noon. We were unleashed upon London, frisky with liberty. A crowd of us ended up at Regent's Park. I can't remember who the others were because, between catching the bus up the Finchley Road and jumping off again, I excised my heart and gave it to Prince C.

He bought me armfuls of daffodils. Skipping up the street, I handed them to smiling strangers. He pretended to be cross. 'What if I bought you diamonds? You'll give them away.' We were wearing school uniform and carrying bags full of books. I'd been dreaming about love since Julie Andrews and Dick van Dyke popped through a pavement picture and danced with penguins, Gilbert Blythe dipped Anne of Green Gables's pigtails in the inkwell and Lulu married Bee Gee Maurice Gibb.

I doodled his name dementedly until the phone finally rang. Would I, he wondered, like to go to the cinema? Gentle Reader, I'd have been ecstatic to accompany him to a sewage farm, leper colony or launderette to watch his socks sloshing in the suds. The cinema? Yes, please. What did I want to see? I was thirteen and four days. I didn't subscribe to *Time Out*. I wasn't au fait with the capital's cinematic bill of fare. Luckily, I'd spotted a poster outside the Finchley Odeon. 'I'd love to see *Towering Inferno*,' I said.

I haven't announced an 'artistic' choice so confidently since. I didn't pause to consider the effect. I hadn't realized films were subject to critical analysis or that my choice would subject me to the same. I was too young. I thought films were lovely. *The Jungle Book*? Lovely. *Hello, Dolly!*? Lovely. *Seven Brides for Seven Brothers*? Lovely. *Those Magnificent Men in Their Flying Machines*? Lovely. Going to the cinema was a treat. I hadn't twigged that I was supposed to have cultivated a worldly-wise ennui. I didn't understand liking some films was more culturally acceptable than liking others. I'd never heard of Jean-Luc Godard or Federico Fellini. I hadn't yet practised shuddering at the mention of Walt Disney.

There was a shocked silence. '*Towering Inferno*?'

'Yes. I'm dying to see it.'

I couldn't sleep or eat for excitement. We met at the cinema and snogged so conscientiously in the back row I barely clocked O. J. Simpson.

On the 26 bus back the deconstruction began: 'What did you think of the film?'

'Fantastic! Great! I loved it, thank you.'

Disgust contorted his features. 'You loved it? Seriously?'

'Yes, it was brilliant.'

Out tumbled a forensic analysis so disdainful it remains etched in my consciousness nearly half a century later. 'The characterization was crude and bristling with stereotypes, the plot leaden and pedestrianly predictable. The suspense was so unsubtly constructed the manipulation was obvious, and don't get me started on the clichéd cinematography.'

I learn fast. I got the hang. The intelligentsia, cognoscenti never say: 'Yeah, fab film. It was great.' They dismantle, compare, contrast, allude to Kurosawa and Ruth Prawer Jhabvala. There are certain etiolated films/novels/poems/plays/TV commercials/documentaries/paintings it's acceptable to admire. The rest are beyond the pale, unless, of course, you like them ironically. I didn't dare pick another movie to snog along to on a Saturday night for nine years.

A teenager in love

How did it feel being a teenager in love? I longed to be called upon to give up my parachute for my beloved or at the very least iron his shirts. I'd have married him and borne his children if the law had permitted. I saw him every day on the morning coach, met him in the cricket pavilion at the boys' school next door for secret snogging and feverish fumbling at break, sat next to him on the coach home and left him reluctantly to hop off the Northern Line at Totteridge station every afternoon.

As soon as *Blue Peter* finished, I ached for the phone to ring. I hadn't seen him for an hour and a half and missed him desperately. We were Romeo and Juliet, Héloïse and Abelard, Heidi and Peter, Barbra Streisand and Omar Sharif. He occupied my every waking thought and every inch of my dreams. Like my father he spouted quotations: 'They hate us youth.' Falstaff, *Henry IV Part 1*.

He approved of Ian Dury and the Blockheads. I immediately memorized all the lyrics on the *New Boots and Panties!!* album. If you fancy booking me to perform 'Billericay Dickie' at your shindig half a century later, my rates are favourable and I'm still word perfect.

Virginity

I won't dwell on the being-divested-of-my-virginity moment, or on my precise age (or lack thereof) at the time. There's no need. I'm a grandma of four, so you've probably guessed my hymen was ruptured quite some time ago. I will say that, if you're going to unload your maidenhead, it's not an altogether terrible idea to do it while wildly in love with someone you're still going to be dementedly enamoured of nine years later. There's nothing wrong with a wriggle with an anonymous stranger but I treasured the romance of true love.

Luckily my parents' obliging GP conjured up period pain as an alibi for putting me on the pill. Sex and love. Love and sex. I'm not sure which was the more absorbing obsession. I was a martyr to my marauding hormones. Sex was electrifyingly alluring because there was a Jewish fatwah on it. As my mother said: 'It [my turning out to be a non-virgin] would kill your father. And we won't make you a white wedding. And we'll hang our heads in shame. And no one will ever marry you, young lady, and you'll end up' – sharp intake of breath – 'on the singles' table at barmitzvahs.'

Prince Charmless was paralysed by angst. I was merry and efficient. Entombed in a perpetual 'essay crisis', he was two terms behind with his schoolwork and teetering on the brink of expulsion. I was handing in perky pages on Shelley's 'Ode to the West Wind' with secretarial promptness. He was suspended in writers'-block aspic and almost hospitalized over the trauma of his O-level French oral. When mine '*s'est produit*' two years later I, distracted by oral of a more licentious nature, had forgotten all about it and, taken unawares, prattled happily about *'randonnée à poney'* (pony trekking). He made a gigantic matza pudding out of minutiae and mundanities. I didn't but contentedly devoted myself to demolishing his demons.

I was thirteen, fourteen, fifteen, ages to be selfish and snog the faces off acne-pitted sixth-formers. Instead, I was worrying about Prince C's geography project (the geysers of Boreham Wood) and consoling him for failing to be cast in the school play *Oh! What a Lovely War* in which I was playing a pierrette. Intent on torturing himself, he accepted the job of stage manager. Pale, wretched and resentful, he cast a pall over my performance and pleasure, and I allowed it. My father did the same to my mother. His torment infected her placid serenity. I was ready and willing to have my equanimity destroyed in the name of true love.

Prince Charmless was destined to read English at Cambridge. Haberdashers' bristled with Oxbridge candidates. In those days winning a place at Oxford or Cambridge usually involved what was called 'seventh-term entry'. That meant taking A levels in the summer and, if you scored a raft of A grades, returning to school in the autumn term when all your classmates had already bowled off to university. Being stuck at your old desk while your contemporaries were having their genitals pierced at Freshers' Week was unutterable hell. You were then hot-housed, till you'd ingested a load of impressive stuff to regurgitate in the garrotting Oxbridge entrance exams in October. The rest of the year was yours to develop a drug habit or a personality, depending on your proclivities. If your chosen college opened its arms to you, you'd backed a winner. If you were rejected, you'd wasted a year of your life.

Prince Charmless wended his tormented, angst-ridden way to Trinity College. I dashed back and forth from Totteridge to Cambridge to partake of his charms. I told my parents I slept in a 'guest lodge'. On the night of his matriculation dinner, he togged up in white tie leaving me waiting in his room. I was going to write 'I was wearing my father's negligée' but that, although technically correct, might give the wrong impression. The garments (nightdress and complementary robe) were oyster silk, slinky, body-skimming, manufactured in the Philippines, as 'come hither' as a welter of synthetic fabrics can be. There was a knock at the door. An unknown 1979's answer to Benedict Cumberbatch whirled in, academic gown akimbo. 'My lord and master! Who have we here? A wanton wench if ever I spied one! Show me your breasts.'

Gentle Reader, to refuse this commanding stranger would have been suburban and stuffy, too terribly Totteridge for words. I obliged. *Quel frisson!* He coolly viewed the vista. We repeated the unveiling on all future encounters, sometimes in the student loos. Prince Charmless didn't twig. It was exceedingly jolly. He instructed. I displayed. He viewed. It was tremendous vintage, doesn't-get-you-pregnant or count-as-being-unfaithful-to-your-teenage-sweetheart fun.

6

'For this you went to Cambridge?'

IN 1979 IT WAS MY turn. I applied to read English literature at Trinity and was offered an interview. My parents drove me there. What to wear that shrieked 'genius' but also 'delightfully brought-up person who reads Seamus Heaney in the shower'? Considering the only one of us to have been to university was my mother, and that was the LSE, the Yiddish chorus had a barrel-load of strong opinions about sartorial choices for interviews at august Cambridge colleges. The consensus was: 'Wear your *yom tov* outfit'. The *yom tov* outfit is the Jewish equivalent of the gentile 'Sunday best'. You wear it to synagogue on Rosh Hashanah and dazzle in it on all momentous occasions for the following twelve months. Mine was a pink pie-crust frilled blouse – I was fashion-forward: Princess Diana didn't wear hers till 1980 – with a woven brown linen skirt and a pink petticoat underlay, so the pink peeped through delicate fissures in the fabric, teamed with pink suede court shoes. I'm not sure the sartorial statement was 'bluestocking meets Virginia Woolf'. I think I projected 'Brent Cross meets Barbara Cartland', but when Professor Adrian Poole asked me: 'If I sent you up to the Wren Library to write a dissertation now, on whom would you write it and why?' I chirped up confidently with a convincing compulsion to churn out fifteen thousand sparky words on (then) undervalued 'Adlestrop' poet Edward Thomas.

The compulsory diet paid dividends. I was a size ten. I had a boyfriend. In December 1979 Trinity College rang. They were delighted to offer me a place. I was relieved. My father burst into hysterical tears and spluttered with sorrow at not being able to tell his own father the news. At this point his father had been dead for nineteen years. So engrossed was my dad in his watery grief, he didn't think to congratulate me, or even acknowledge the emotional deluge had anything to do with me. This was another Norman-Feltz-monopolized moment. His emotions engulfed everything. He was fatherless, suffering, in torment. I was irrelevant.

My mother's reaction was a tad niche: 'You know you'll fill up on chocolates!' What? Not even a '*Mazeltov!*' after all those years poring over Milton, Virgil and Horace? Not even a small, understated 'Well done'?

Mother was right. I filled up on chocolates

Mother knows best. Mine was spot on. One of the most enchanting elements of student life at Cambridge was, just as she predicted, freedom at last from her dietary tyranny. She wasn't there and couldn't put a lock on my biscuit tin. Sainsbury's was steps away. I had a whole fourteen pounds a week to spend on Food! Glorious Food! Naturally, I chose formerly forbidden delights: apple juice, Nutella, Feast ice creams, bread thickly spread with butter and jam. It was as if wartime rationing had come to an end, and I raided a limitless larder to make restitution to myself for years of deprivation.

When I went home at the end of term my mother said: 'Shall I bounce you in? You're a great big pumped-up beach ball.'

Whatever happened, I wonder now, to 'Welcome home'? Was she right to worry? You decide. I was a size ten to twelve, curvaceous, bosomy, possibly edging towards blowsy. I wasn't fat. I was fit, fanciable and a normal size for a girl from a dumpling-shaped family. I was more Pekinese than whippet but that was on the genetic cards. There was no cause for caustic criticism or counting each individual cashew nut I consumed.

My mother, my dealer

By the time I was twenty my mother quietly began scoring diet pills, amphetamines, from her hairdresser and passing them to me. There's nothing quite like taking drugs illegally obtained by your loving mum. They were a spectacular success. Stellar stuff! The pounds magically evaporated. Flesh melted into oblivion. The side effects were nightmarish. Revved up on speed, your heart races like the thumping overture-and-beginners to a fatal cardiac arrest. You can't sleep. Nights are a sweating, churning, pulse-throbbing orgy of insomnia. The thought of food makes you so nauseous you heave. You are chronically ketotic. Your breath reeks of nail-varnish remover. Everything is hectic, hallucinatory and garish.

Mum's helpful donation coincided with my Tripos finals. The exam papers were a hieroglyphic blur. Who cared? Chunks of skeleton were visible beneath my skin. I wasn't just slim but that holiest of holy grails – thin, verging on gaunt. Dark circles rimmed my eyes. I looked like a malnourished bushbaby. I loved it. It was all I'd ever wanted. My mother was beside herself. She gave the hairdresser a thwacking tip. 'Get us lots more, sweetheart – pronto.'

Valerie was high as a kite. She'd sampled the goods, adored the effect and was wowing the neighbourhood in skintight jeans, the slimmest at forty-three that she'd been, she said, since her eighteenth birthday. I felt sick. I needed to puke. I slid into a size eight. I was exhilarated. I wanted to chop my head off. I sprouted hip bones and clavicles. Amphetamines! What a find! A miracle cure – except the instant we stopped pill-popping the pounds piled straight back on and then some.

I didn't eat a morsel of the formidable feast concocted by my mother for my twenty-first-birthday party. For days we whipped, frothed, iced and basted. I was the galley slave. She was Nigella. My mother excelled at fancy fare. Her marron mousse had to be inhaled to be believed. My famished, freshly graduated friends, who'd been drinking for days and subsisting on nothing more solid than Hula Hoops, tucked into the ambrosial manna chez Feltz with gusto and gratitude. Simon Tuke disappeared under the piano with a cut-glass bowl containing a whole family-sized Valerie Feltz extravaganza trifle. Andrew Shelley persuaded my mother to conjure up another spontaneous soufflé at four a.m., so enamoured was he of the three he'd already swallowed. The food was celestial, exotically flavoured nectar. Bouches were amusés. I, the Birthday Girl, didn't lick a spoon or dip the tip of my finger into the vichysoisse. I've regretted not sampling my own birthday banquet thousands of times over the years. I couldn't do it on the night. I'd ramped up my amphetamine dose and the little yellow pills wouldn't let me.

Cambridge was just as you imagine it to have been in 1980–83. Lots of posh people strode about in Barbours, jodhpurs and green wellies, braying about being 'so pissed, I puked on a porter'. The few state-schoolers, brightest in their school/village/conurbation, staggered stricken and wrongly dressed into supervisions where indolent Etonians vanquished them with insouciant bons mots. Jews didn't fit in. There wasn't a pigeonhole for wandering pedlars. Some took one look at the stained-glass windows and pretended not to be Jewish – my dad warned me about crypto-Jews – doing their utmost to regrow foreskins and learn the Lord's Prayer.

I was a private-school-educated, cut-glass-accented Jewess. That made me confusing and exotic. My *yom tov* outfit sent mixed messages. Was I more town than gown? Was I rich/ridiculously formal/Mata Hari? I didn't pass unnoticed, yet some didn't know quite what they were noticing.

Evelyn Waugh's novel *Brideshead Revisited* was elegantly adapted for television in 1981, starring chiselled-cheeked posh boys Jeremy Irons and Anthony Andrews. The series enthralled plebs and patricians alike. Some impressionable undergraduates combed their nurseries for abandoned teddy bears and dedicated themselves to wafting about the Backs in brocade dressing gowns à la Sebastian Flyte.

My first essay, set by Professor John Marenbon, asked: 'What did Chaucer do to "Il Filostrato"?' Hang on. What? *Il Filostrato* was a confection of Boccaccio. Once I'd penetrated the classical Italian, I twigged that in his throbbing *Troilus and Criseyde*, Chaucer turned Boccaccio's cardboard ciphers into full-throttled characters from *Coronation Street*. What can I say? I'm a swot.

I was too shy to venture through the hallowed portals of Footlights, the legendary theatre company that incubated Stephen Fry, Hugh Laurie and Emma Thompson and hatched them into instant stardom. Instead, I chose student journalism. Hiding behind the nom de plume Vivacious Loyola – in 1981 we undergraduates made puns about sixteenth-century theologians: that's just how we rolled – I had a rollicking time reviewing theatrical productions. In May 2024, I bumped into former *Evening Standard* literary editor David Sexton. He texted me the morning after: 'I remember you reviewed Tilda Swinton nude in an Athol Fugard play and said it answered the question of whether she was a natural redhead. Way to go.'

After university I didn't waft about with the cognoscenti. I might have spotted some of them draped in Pre-Raphaelite garb over an ornamental gargoyle in Great Court, but I hadn't forged any lasting influential liaisons. I'd made lifelong pals, but not the kind with literary salons in Belgravia. I'd done my washing in the same launderette as Emma Thompson, eaten a Fitzbillies bun in an adjacent punt to Stephen Fry's and seen Hugh Laurie taking a tactical chunder in a gutter on King's Parade but I wasn't up close and personal. Scroll forward nearly ten years to 1992 and watching the film *Peter's Friends* directed by Kenneth Branagh and starring Stephen, Emma, Hugh and Imelda Staunton would have been bliss if my mother – yes, I was thirty and still went to the cinema with my parents – hadn't lamented in a stage whisper: 'Oh, Vanessa! Why weren't you one of Peter's friends?'

I was punted to Grantchester by a lovelorn swain who played me twenties ditties on a wind-up gramophone. Prince Charmless loomed around for the first year but women were in such short supply a boyfriend was a minor impediment to declarations of adoration from besotted suitors. I had a ball, often at a ball. Prince Charmless's charms charmed me less. His fifth-form essay crisis expanded into a real-life 'Why have "they" not snapped me up to be director of the National Theatre?' crisis. One evening while Charmless was mid-moan, a terrifying vision of a future spent listening to a trust-funded, full-time whinger bewailing the world's idiocy for not putting him in charge of the Royal Shakespeare Company shook me to the innards. I'd been listening to the litany since my thirteenth birthday. I'd had enough.

Farewell, Prince Charmless

PS Prince Charmless might have sashayed sweetly into this book enshrined for ever in the romantic first-love bracket, had he not phoned out of the blue twenty-five years later and asked me for three thousand pounds. He didn't trouble to ask politely. Somehow, he managed to be arrogant and boastful and demand money all at the same time, thoroughly nuking the soft-focus teenage prism in which I'd consecrated him. I was in a taxi. The driver turned round (literally) and said: 'V, don't you dare!' He needn't have worried. Prince Charmless's parents were Monaco-based tax exiles. I had no intention of raiding the Bank of Feltz.

7

'Darling, you have delighted us long enough'

1983: I GRADUATED AND CAME home. My parents were appalled. I had wasted my time, squandered my opportunities. How, they wondered aloud, had I managed to spend three years heavy petting in punts with eligible chaps and failed to secure a single proposal of marriage?

My father, paraphrasing Jane Austen's Mr Bennet, declared: 'Darling, you have delighted us long enough.' He meant: 'Your time here is up. By all that's holy, hurry up and get married, preferably by this afternoon.'

I could have answered: 'Hang on! This is my home. You're my dad. I'm only twenty-one. I've been studying solidly since birth. I've seen and done almost nothing. I don't want to wash a man's socks. This is 1983, for fuck's sake. Get off my case. Stop harassing and haranguing. Shut up.' But I didn't. No one I knew would have dreamed of speaking to their father like that. I didn't even dare think it. I meekly accepted the diktat. You're not welcome here. You're single. You're an embarrassment. The clock is ticking.

All that 'groves of academe' bright-young-thing stuff was packed away neatly in the loft with my university kettle and rape alarm. My father said I had six months to make a living as a journalist or I'd have to requalify as a solicitor. I didn't want to be a solicitor. He was not interested in my preferences. Six months isn't long. I went to WHSmith in Golders Green and scribbled down women's magazine editors' phone numbers on the back of a bus ticket. I rang them all. Some told me to get stuffed. Others were prepared to give a graduate novice a try. I scratched out quizzes and features for any magazine that would cough up, among them insightful aperçus on breast-feeding, which I knew nothing about, for *Mother & Baby*.

I was a disappointment to everyone – especially myself: surplus to requirements at home, my youthful promise already evaporated. I'd peaked at twenty-one. Cambridge had been the pinnacle, and I was sliding snakes-no-ladders-style down life's board.

Lorraine Davidson next door got engaged. I was invited to view the bride in her wedding gown. She was luminously ethereal in embroidered satin. Stumbling the few yards home, I was crying so hard I couldn't see my own house. More than anything else on earth, I wanted to be her.

Just so you're in no doubt, if I didn't get married, I couldn't do anything else either. It's not as if I could go and share a flat in Shoreditch with a girlfriend. I couldn't waft off to a yurt in Somerset or hang out in a commune in Marrakesh. I wasn't allowed to do anything except work and wait. I was powerless and penniless – a resoundingly rubbish combo.

8

Grandma Sybil sends me to Casualty

1984: GRANDMA SYBIL WAS IN University College Hospital dying of leukaemia. A doctor came in to take a blood sample. 'Doctor,' said my grandma, 'are you Jewish?'

'Yes.'

'Are you married?'

'No.'

'Have I got a girl for you!'

I was twenty-two. The spinster thing was an embarrassment for the entire family. Grandma Sybil was determined to shove me off the shelf 'if it's the last thing I do'. I bowled into her room brandishing a bunch of grapes. 'This is no time for greengrocery. Brush your hair. Spritz yourself with my Diorissimo. Put on some lipstick. Take yourself down to Casualty and introduce yourself to a doctor I met.'

I don't know if I've mentioned, for Jewish people, a stethoscope is a sceptre. Doctors are royalty. Doctor worship is part of our mythology. The theme runs vividly through Jewish humour. Mother on the beach: 'Help! Help! SOS! My son the doctor is drowning!'

There was nothing humorous about Grandma Sybil's determination. I tried to stall. 'Grandma, you can't be serious. You want me to walk into A and E and accost a strange doctor?'

'There's nothing strange about him, Vanessa. Am I your grandma? Do I know what's best for you? All right, now go.'

Gentle Reader, I went.

The department was foaming with blood, pus and crying babies. 'I've come to see the Good Doctor. My grandma sent me,' I said, to a rushed-off-her-feet nurse. I used to tell this story all the time. It was my own personal fairy tale.

Just for embellishment, I'd say the nurse replied, 'Oh, God! Not another one.' She didn't really. She just trotted off and minutes later a twenty-five-year-old Mancunian with wavy black hair, startling blue eyes and a dimple in his chin appeared.

'My grandma sent me,' I said, doing my utmost to ooze va-va-voom to expunge the embarrassment. He scribbled my parents' (landline) number on the collar of his white lab coat in biro.

When I told Grandma Sybil, she was furious: 'You should have written it down for him on a piece of paper. He'll send that coat to the hospital laundry. You'll never hear from him.'

He rang, asked me out and borrowed a car from a friend. He was serious and earnest. We walked past a minor car bash. He dashed across the road to see if the bopped driver needed triaging. Never underestimate the allure of a chap who can ace the Heimlich manoeuvre and diagnose glandular fever. Knowing he could give me the kiss of life made me feel like kissing him. He was sincere, not sophisticated, resourceful, not rhetorical and absorbed by medicine. As the daughter of a self-dubbed tragic hero who blamed the universe for locking him in a warehouse full of brushed-nylon pyjamas, it was bliss to be with someone comfortable in their calling. He'd travelled to Papua New Guinea and up the Amazon, making my feeble forays to Marbella and Bournemouth seem pathetic.

How my heart was won

If you're wondering how he won my heart, don't. I'll tell you. He asked what I did for a living. I told him enthusiastically I'd just finished a piece for *Majesty* magazine on Princess Diana's maternity wardrobe. (She was pregnant with Prince Harry.)

He was disappointed. 'Isn't that superficial? Don't you feel you should be covering something worthwhile? What a waste of your intelligence. You must only be using half your brain.'

Please don't tell me what I should have said. I know. I ought to have replied: 'How judgemental and condescending you are. Drop me off at the nearest tube. I'm out.' I didn't, of course. You probably know why. His disapproval was warm, familiar territory. My parents were scornful of and disappointed by – I was going to put 'my choices' but *le mot juste* is 'me'. I perpetually let them down. They were disgusted I hadn't had 'a Communist phase', appalled I'd preferred *Bouquet of Barbed Wire* to *Anna Karenina* and when I was fourteen 'aghast' I 'still' hadn't read *Silas Marner*. I was used to being rebuked for my shallow lack of intellectual rigour. They wanted Susan Sontag, American writer, critic and 'public intellectual'. They got me, playing Liza with a Z and reading *Tatler*. An unknown doctor concluding within minutes of meeting me that I was selling my intellect, family and society short by focusing on fripperies instead of writing about inner-city housing decay was exactly what I thought I deserved. He gets it! I thought. He sees through me. He knows I should be doing something worthwhile. He is saving lives. I'm debating taffeta over organza. Wow! The evening ended with a gratifying flurry of French kisses.

I interview George Michael and Andrew Ridgeley and wish George had been that kind of boy

A few days later, I swung by the Good Doctor's digs after interviewing George Michael and Andrew Ridgeley. Wham! had just released 'Wake Me Up Before You Go-Go'. Their family and friends had been ringing the Capital Radio hit-line all weekend trying to propel it to number one. Mid-interview the Capital Hit Parade was broadcast live. Would I mind very much if they listened? We listened. They chewed their fingernails. Wham! made it to the number-one slot. The boys – they were boys, both twenty-one – were electrified. They grabbed me and started dancing, first on the floor, then on the sofa.

George was gorgeous. If I'd been that kind of girl and George had been that kind of boy, I definitely would have. He wasn't. We didn't. I wished we had. I still wish we had. He was lovely to me whenever we collided ever after. When George became tormented and unhappy, I tried to remember him that spring day in 1984 in his agent Connie Filippello's office when he was young, happy, hoofing on the sofa, on top of the world.

9

Quick! Book the caterers!

THE GOOD DOCTOR WAS OUT. I left a note. I'd just joined the Thompson Twins on the road writing about their tour for *19* magazine and 'Doctor! Doctor!' was their super-hit. 'You are out. Shame! I came to ravish you!'

He had a spag-bol party. He could cook. I washed up next day while he was saving lives. He had one jacket, terrible shoes, a jumper he'd picked up in the Himalayas that should have stayed in the Himalayas, no car and was earning £6,750 a year. I couldn't drive, was back in my childhood bedroom and was probably earning around three thousand. I loved him. He loved me.

The romance reverberated against a backdrop of mortality. A child was rushed to Casualty. His parents held a crisp packet containing his fingers. He'd stuck his hand into the monkey enclosure at London Zoo. The Good Doctor was on duty. A UCH consultant sewed the severed digits back on.

Grandma Sybil was deteriorating. She was sixty-nine, the fulcrum of the family. Poppa Willie, told by her haematologist his wife was not long for this world, chose to hear that she needed a nourishing diet to recover and tried to persuade her to drink eggnog. If humans want to badly enough, we hear what we wish to and erase what is unbearable. I zigzagged hourly from true love to grief, from walking hand in hand with my beloved to watching helpless as porters scooped my skeletal grandma on to a trolley to subject her to yet another painful – ultimately pointless – procedure.

Meeting the parents

If you've seen *Meet the Fockers* you'll have an inkling of the atmosphere when I was taken to Altrincham, Cheshire, to meet the Good Doctor's parents. We'd known each other ten weeks. His father warned him, 'There is such a thing as drowning in honey. It's sweet, but you are still drowning.' Late on the Saturday night, the Good Doctor asked me to marry him. Of course I said 'Yes!'

At nine a.m., he phoned my father to ask for my hand in marriage. Talk about a formality! My mother was already on the phone to the caterers, a handful of pastel swatches for the bridesmaids' dresses in her bag. They leaped into the Bentley (fourth-hand but his family didn't know that) and made it to Manchester by lunchtime. The Good Doctor's mother whipped up her notorious salmon mousse. The Good Doctor's father capered about the drive photographing the moment for posterity.

We were all grinning so broadly our faces ached. We didn't care. What was a spot of facial strain when it was finally happening? A wedding. A *simcha* (celebration). A married daughter. All those school fees. All those prayers. All that waiting, wishing and hoping. Thank God, *Baruch Hashem* (blessed be His name), an end to the ignominy. Bye-bye to the singles' table at barmitzvahs. Hello, grandchildren. At last, I'd managed a coup they could be proud of. The Good Doctor and I had known each other only ten weeks. No one breathed a syllable about 'taking our time'. Getting to know each other better was an alien concept. He'd asked. I'd accepted. What more did anyone want? Put a spoke in the wheels by suggesting a long engagement? Don't be ridiculous. You looked. You found. Get on with it.

The wedding was booked for 31 March ten months later. The engagement party – of sorts, Grandma Sybil was so ill no one had the heart for a big affair with dancing – was booked and paid for before a mouthful of mousse had been swallowed. My mother asked the Good Doctor: 'Is there anything in particular you'd like to eat at the party?'

'Coleslaw,' he blithely replied. She gave him a look of withering disgust. The word 'coleslaw' was never uttered again.

I was so happy you could have warmed your hands on my heart. I was radiant. I was so sad. I was in despair. I'd found the man I adored. He loved me right back. The grandma who presided over all our lives was dying. How would we be us without her? The guilt was intense. It isn't seemly to be so happy when your grandma is a wraith, hollow with pain. It isn't appropriate to be so sad when people are pinching your cheeks, saying, 'Come here, *callah* [bride],' and buying you twelve wine glasses and a matching decanter.

Grandma Sybil died in August 1984. Jewish funerals are stark. Rich people can't buy shinier, more glamorous burials than poor. There are no flowers. I mean it. Not a bloom. There's no music, no horses, no processions. Our coffins look as if they're made from balsa wood and matchboxes. You can't pick from a glossy menu of walnut lined with pale pink satin. All are covered with identical cheap black tarpaulin. Ashes to ashes. Dust to dust. No gold corbels. The *levoyer* (burial) takes place within hours of death. It's a mad dash, skin of the teeth, edge of the seat. Special measures must be taken to bypass the bureaucracy and vault the usual snail's pace of British funerals. The official go-ahead is so last-minute people are still phoning around with the information as the mourners depart for the ceremony.

We don't wear black. That would be *hukkot ha goyim* (adopting the custom of non-Jews). But we are British. We know no other country. Wearing any colour but black to a funeral feels all wrong. Everything feels all wrong. Mirrors are covered, low chairs are delivered. Aunts and cousins converge on the kitchen. The Good Doctor, Prince Charmless and all my friends were at Bushey Cemetery. The customary greeting at funerals and *shivas*, the seven days of formal mourning that follow, is: 'Wish you long life.' I lost count of the people who embraced me with a joyful: '*Mazeltov!*' blushed, turned their smiles upside down and said consolingly, 'Wish you long life.'

Your wedding? Frankly, you're lucky to be invited!

We had a wedding to plan. I say 'we', but my parents made it clear from Engagement Day One that it was 'their wedding'. This is what they said: 'We have been going to friends' children's weddings for years. This is our chance to reciprocate. You may have one table of guests. The Good Doctor may have one table of guests. We'll fight it out with the *mechatunim* [in-laws] over how many tables we decide they can have. That is a matter for adults – nothing to do with you. This is our wedding.' I wouldn't have been in the least surprised if they'd added: 'You're lucky to be invited at all.' They were paying and the 'astronomical' cost was bewailed on the hour every hour. They called the shots.

My presence was barely required at any point except, fairly late, on the day itself. I wasn't taken to the florist to discuss the flowers. I had no say in the menu, the choice of caterers, the table plan or the band. Great Portland Street Synagogue, London W1, where my parents tied the knot, and the Royal Lancaster Hotel ballroom were booked. The Good Doctor and I were permitted to visit both to gaze at the fait accompli. I was allowed to express a preference for pink and to ask for ten bridesmaids. I knew how much I'd loved my seven sallies up the aisle holding trains and posies and wanted to be trailed by as many excited little cousins and neighbours as I could recruit.

My dress, from Andrea Wilkin in St Christopher's Place, was a dead ringer for the cake. Both were crisp white crinolines, adorned with pale pink roses. I was a frantically dieting Scarlett O'Hara. White satin shoes with tiny pink silk roses were ordered. My mother wed in a short veil. I was commanded to choose likewise. Hourly she wondered aloud: 'All brides lose weight before their weddings. When, oh, when is it going to happen to you, Vanessa?'

Searching for a headdress or tiara to wear with the veil failed. My mother's response to everything I tried on was a despairing 'No! Absolutely not. Why do you keep doing that thing with your face?'

What thing, Gentle Reader? I wasn't 'doing' anything with my face. That was my face. I couldn't do anything to change it. My mother decided there wasn't a headdress on earth with the razzle-dazzle to counter that 'thing' I was doing with my face, so there was no point buying one. What was the purpose with a face like mine? The veil would just have to be pinned directly to my hair.

Born-again blonde

While we're discussing hair, mine had been bleached blonde every three weeks since I was seventeen. Inspired by Auntie Pat Layton round the corner and Diana Dors on the cover of Uncle Melvyn's *Sgt. Pepper* album, I am and have always been a born-again blonde. My roots may be dark – each time they grow out, I am horrified – but my soul is blonde. The Good Doctor didn't like it. He'd fallen in love with a blonde, but the moment we were betrothed he declared his dislike of the 'brassiness'. I tried to explain that was the whole point. I wasn't trying to look natural. My aim was the opposite. He wasn't convinced.

I went brunette. If your husband-to-be doesn't like you blonde, you stop being blonde. I bumped into former snogging partner Michael Segen at the Water Margin Chinese Restaurant (of blessed memory) in Golders Green. 'What the hell has happened to your hair? It makes your face look green.' Brunette didn't suit me, spiritually or physically. I was a brunette bride.

The morning post brought pre-stamped cards accepting or declining the invitation. My father was elated and dejected with every spoonful of Weetabix. 'The Goldbergs can't come. Thank God. I can't stand them. They've enclosed a gift. A cheque. Fifty pounds. Not bad for no-shows. Valerie, who are the Lipschitz? Cousins of the Good Doctor? *Oy a klug!* [Woe is me!] There are five of them. They're all coming. *Oy gevalt!* [Woe is me!] What do they think I am? Made of money?' The avocado-shaped avocado dishes piled up. We scored a goodly smattering of oven-to-tableware, fondue sets, a TV from Uncle Sam Beckman and a collection of linen duvet covers and pillow sets requiring hours of ironing.

10

Going to the synagogue and I'm going to get married

MY WEDDING DAY, 31 MARCH 1985, was sunshine-dappled and dizzyingly delightful. My father took me for a walk on Totteridge Common to stay my nerves. My mother made sure I went braless. 'You don't want the straps to leave a mark when your wedding dress is off the shoulder.' I was too happy to breathe. I didn't need oxygen. I was manufacturing my own. I was a bride. Hair and Make-up had arrived to make me look like one. I had petticoats sticking out so I had to be lifted over them to step into the middle of the hooped multilayered net skirts. My dress was snowy silk with a boned bodice, the crinoline scooped into scallops and scattered with pale pink silk roses. When I appeared, my ten little bridesmaids all in pink hopped and skipped with excitement. My father handed me down the stairs, sobbing convulsively.

My mother had invited all the neighbours who weren't coming to the wedding for drinks and vol-au-vents in the drawing room. I entered to a rousing round of applause. Was I being clapped for looking beautiful, or for finally landing a husband? Frankly, it didn't matter any more. *Sine qua non* of wedding photographers Charles Green captured every nuance on newfangled video.

My mother can clearly be seen and heard shushing me

throughout. Whenever I exclaim about how happy I am, she responds: 'Vanessa, be quiet.' I was quite quiet. Jewish brides are silent. You don't get to say, 'I do.' I wasn't asked to make a speech. Most of our 250 guests would come, eat poussin, dance and go without hearing a syllable from me.

Brides being driven to church with their fathers is a British custom. Jewish brides travel with their mothers and mothers-in-law. My mother-in-law was so appalled by the ostentation of the white wedding Roller festooned with pale pink ribbon, she shrank in horror to the size of a hamster and spent the whole journey hanging her head in shame. On the corner of Portland Place the late Kenneth Williams, crossing the road with his mum, waved and blew me a kiss.

My groom was attended by twenty-three men in top hats, white tie and tails. My mother thought a common tux was . . . common. A phalanx of ushers, uncles, cousins wafted me up the synagogue steps. I was shown into an anteroom to wait for the *bedecken* – the moment when the groom, to avoid being tricked as Jacob was into marrying ugly Leah instead of beautiful Rachel, lifts the bride's veil to check he's not being palmed off.

None of us had met the rabbi. His name was Rabbi Shine. He left quite an impression. Watching the video thirty-nine years later, I am shocked afresh. The Good Doctor, trembling, walks in to inspect his bride. Lifting my veil, he bursts into tears, saying, 'Precious! You look so beautiful.' Into the idyll plonks Rabbi Shine. With the musicality of a foghorn, he bellows, in a middle-European accent: 'Don't get too busy now!' Busy? Bathed in love at our own wedding? Perish the thought.

Gliding up the aisle on my father's arm was the apotheosis of my existence. This was the moment I was born for. This was my purpose. Towering flower arrangements scented the synagogue. Giant pink gerberas swayed on fragile stems. Our *chuppah* was a canopy of fragrant blooms. Why was my bouquet shaking so violently? I tried to still it with a two-handed grip. I couldn't. My posy was shaking because I was shaking. My mother, in turquoise silk with bulbous puffed sleeves and an ingenious spangled jacket that could be removed at dinner to reveal her décolletage, seemed to be weaving up and down. So did Rabbi Shine. I'd never fainted. This was not the moment to start. We'd agreed in advance I would weave a circle seven times around my groom to teach that evil eye a proper lesson. I encircled the Good Doctor twice before he told me to stop. I was doing exactly what I'd been told to do but it seemed to irritate him. I stopped circling. I didn't complete the preventive ritual. Maybe that gave the *ayin hara* (evil eye) a chink to slip through.

Wedding highlights

I'll give you highlights. The Good Doctor pimped his white tie. All twenty-three chaps wore identical white waistcoats and hand-tied white bow ties. His were pale pink silk. He'd been to Andrea Wilkin in secret and commissioned her to make his especially to complement my dress. I was touched beyond measure.

At the synagogue, the Good Doctor's brother guided guests so zealously into cars to the reception at the Royal Lancaster he was left behind in a rented suit without a penny to his name and had to trek on foot the whole way to Hyde Park, turning up exhausted and windblown in the middle of the hors d'oeuvres. On the receiving line, the Good Doctor was in a rhythm, kiss the lady, shake hands with the gentleman. Kiss the lady. Shake hands with the gentleman. He missed a beat, shook hands with Auntie Freda and gave Uncle Mick a passionate smacker. Uncle Mick was surprised but flattered. The Good Doctor whispered he didn't care if

Mick was Grandma Babs's brother, he was much too embarrassed ever to see him again.

Seated at the top table, too scared of dripping gravy on my virginal white silk to eat, I read the individually printed menu. The dessert: *'Délices à l'ananas Marjorie'*. 'Who is Marjorie?' I asked my mother.

'For God's sake, Vanessa, stop making such a fuss. I'm dealing with five previously undeclared vegetarians.' I wasn't making a fuss. The wording was supposed to be: *'Délices à l'ananas Vanessa'*. It was my wedding.

Dinner was interminable – four courses, the cost of each sending my father 'to the poorhouse'. My father's proposal of my health was an encomium to his own multitudinous achievements. He didn't stop the flow by mentioning me. The Good Doctor rose to his feet. The band struck up. It turned out he'd been rehearsing with them in secret for weeks. He serenaded me to the tune of 'On The Street Where You Live' from *My Fair Lady*. 'Just ten months ago, I was all alone. I just wanted someone I could really call my own. You came and found me, down in CASUALTY, when you knocked on the door to my life.' There wasn't a dry eye in the house.

I was aloft, defying gravity, floating on cloud nine, looking down at the party, beaming. My husband. My angel with perfect pitch, perfectly pitched at me. I couldn't wait to start our life together on the third floor of a mansion block with no lift overlooking the Hendon Way. First, though, there was the minor matter of surviving the honeymoon alive.

11

A little bit of dysentery for you, darling, a little bit for me

YOU'RE WONDERING WHY WE CHOSE a risky honeymoon? We didn't. There was no white-water rafting or danger of being munched as we slept by marauding wildebeest. We went for a fortnight to a four-star hotel in the Benidorm of Israel, Eilat, the King Solomon's Palace. The Palace was four-star because there was no room service. Relax. This is not too much information. I am arming you only with facts of the utmost importance. A further important fact is my mother's pre-wedding-night advice. 'Preserve your mystique. Never at any time in my twenty-five years with your father have I allowed him to see me shave my legs, much less see or hear me . . . ! Mystique is your primary weapon. If you don't preserve it at all costs, I cannot vouch for the consequences, Vanessa.'

Our honeymoon coincided with Passover, the festival marking the Israelites' escape from slavery in Egypt. So hastily did our ancestors flee servitude when Moses divided the Red Sea, they didn't wait for the bread they were baking to rise. On the eight days of Pesach, we eat only matza, constipating cardboard, unleavened bread, to remind us of their travails. The hotel lobby housed chill cabinets with glass doors. Inside were plates, each with two slices of matza and two ice-cream scoops variously of egg mayonnaise, tuna mayonnaise and salmon mayonnaise. Peckish guests would help themselves, trying to snazz up the penal matza by loading it liberally with savoury spread.

We newly-weds shared a portion of matza plus egg mayo. I fed him. He fed me. Indulge us. We'd only been married a matter of hours. 'A little bit of egg mayo for you, sweetheart.'

'A little bit of egg mayo for you, precious.' We divided the snack scrupulously into equal halves until there wasn't a scrap of oleaginous gloop left on the plate. Within hours we were stricken. We were burning with fever. We were too ill to walk, talk or read. We were spewing and spouting from both ends simultaneously. There were gut-wrenching sound effects as our guts were wrenching.

I was trying to preserve my mystique but couldn't sing 'La Marseillaise' to cover the noise and projectile vomit at the same time. This is where the 'no room service' bit becomes relevant. We needed water. They wouldn't bring us any. 'We are too weak to get to the lift. We can't get it ourselves,' we whispered.

'You should have gone to a five-star!' they snapped back. We were dead ill. We got worse. The pictures of the wedding were still being developed and we were – I'm not exaggerating – crawling along the floor to the bathroom and back in relays. An occasional puff of stale air blew from the air-conditioning unit. I was sure we were dying of legionnaires disease.

We thought we were the only sufferers until there was a military-style hammering on the door. We were told there'd been an outbreak of shigella dysentery in the hotel. An infected sous-chef had contaminated the mayonnaise. Oh, my God. Not the mayo. Not those enticing blobs of egg, tuna and salmon glistening tantalizingly with lubricious mayonnaise. We were far too sick to laugh. This wasn't the time for gallows humour. Babies were in intensive care. Old people were being carried out on stretchers. The hotel was quarantined. The man was talking but I couldn't concentrate. I wasn't well enough. My stomach cramps were excruciating. The doctor, gold teeth, gold watch, took a look. 'She needs to be admitted to hospital immediately. We will take out her appendix. She has a-ppen-di-ci-tis!' A-ppen-di-ci-tis!

I turned to my husband the doctor. 'Appendicitis? Do you think so? Do I need an operation?' His skin was green. The whites of his sunken eyes were ochre. He phoned my father. 'Norman. We are ill.'

'Ill? What are you talking about? What's the weather like there?'

'No, Norman, we are sick. There's an outbreak of dysentery. There's a doctor here talking about taking Vanessa's appendix out and I don't know what to do.'

'Do?' said my father. 'Do? *You* are a doctor. I am in the underwear business. How the hell do I know what you should do? This is a long-distance call. Don't waste your money.'

We were half carried, half bundled into the lift, shoved into a van and deposited at an army field hospital in the desert. Conditions were brutal. The place was flooded with mayonnaise casualties. Take the word 'flooded' literally. Spurting was occurring from all orifices. There was a deficit of British stiff upper lip. High-decibel keening, wailing and puking from the international invalids immolated within added to the misery. A scary military medic appeared lugging a drip attached to a primitive wooden frame on wheels. He flicked vigorously at my sunken skin trying to locate a vein. At that moment, the Good Doctor staged a dramatic recovery. As if by divine intervention, he morphed from wan, comatose sick person into healthy recoveree fit to run a marathon. He didn't fancy having his veins perforated with one of their prehistoric needles. I had no choice. My new husband wished me well and decamped to the hotel. I was left, like all the other inmates, steering my unwieldy drip contraption to and from the only lavatory being shared by a hundred seeping dysentarians. I won't affront your sensitivities with details of stench and eschatological spatterings. You don't deserve it.

12

First comes love, then comes marriage, then comes Vanessa with a baby carriage

My parents picked us up at Heathrow. For once, having been nil by mouth for ten days, I was almost thin enough to satisfy my mother. 'Let's just hope you don't squander this opportunity by putting everything you've lost straight back on again,' she said, looking worried. Hooray! Married life was launched. Here's the fantastic thing. Until the moment my husband walked out the door fourteen and a half years later, I was blissfully happily married. Obviously, I wouldn't have been if I hadn't believed he was too. I loved hanging out with him. I loved putting my Sabbath tablecloth in the washing-machine. I loved inviting friends over for M&S salmon en croûte – one of the first ever ready meals and all the rage in 1985.

Within three months, I was pregnant and ecstatic. We had minus money. We were clever but pathetically low earners. I was bashing out articles. He was curing the halt and the lame. We were both poorly paid, and our mortgage soaked up almost all our joint earnings. Life wasn't in the tiniest particle glamorous. I was twenty-three and staring at the traffic on the teeming Hendon Way out of my third-floor window. The *Saltburn* vibe of Cambridge was a universe away. I was jolly, chipper and loving life.

Allegra Sybilla

A no-frills birth on the blessed NHS at London's University College Hospital was a touch basic but gloriously productive. Having my waters broken with what looked like a crochet hook in a public holding area in which the woman-next-to-me's partner was eating a Big Mac wasn't ideal, but it was worth it. The prize was exquisite seven-pound-eight-ounce Allegra Sybilla – named for my grandma – the most gifted and beautiful baby in the history of going forth and multiplying. It was adoration at first sight. It still is.

I mugged up on the baby-rearing manuals by doctors Penelope Leach and Miriam Stoppard as if I was revising for Cambridge finals. There wasn't a sentence in Professor Gordon Bourne's *Comprehensive Guide to Pregnancy, Labour and Motherhood* I hadn't underlined in crimson ink. Of course Allegra hadn't read them, so she developed in her own ineffably sweet way, which bore no resemblance whatsoever to their 'expert' descriptions. I had digested so much guff about bonding issues and conflicting mother–baby personalities – but one stuck in my mind: 'You don't like every adult you meet. Why should you like every baby?' I worried. I needn't have. It's a cliché until it happens. I was suffused with unconditional love for my daughter. If she sneezed, I applauded. She was, and still is, the perfect child for me. I've never once wanted to send her back to the manufacturers and exchange her for another with different buttons. When she fell asleep, which happened rarely, I immediately wanted to wake her up because I missed her company so much.

Cambridge. Marriage to a Jewish doctor. A flat. A baby. I was only twenty-four years and three weeks old. Hadn't I delivered on the parental requirement to the letter? OK. I hadn't yet won the Nobel Prize, but there was still time. Was it too much to pray that, just occasionally, Mum and Dad might wallow in the *nachus* (pride in your child's achievements) and pipe down about my weight? I kept hoping they'd see beyond my size and join, with gusto, in the life I had made for myself, situated conveniently just round the corner so they could share every second without a schlep.

At least once a week, when my parents came to dinner, I'd blow-dry my hair and wear my prettiest dress. My flat would smell inviting. I'd roast a chicken to a turn. The bell would ring. Valerie and Norman would walk in through the door. They'd see me ... and dire dismay would cloud their faces: 'Vanessa, you could be so pretty if you just lost weight. Why are you doing this to yourself? We saw Caroline Rose today at the golf club. She was looking gorgeous. You could look like that. If you lose a stone, we'll buy you a dress/send you to a health farm.'

Remember, I didn't do the British thing and see my parents once a year at Christmas. I saw them all the time, every week, multiple days a week, sometimes several times a day. At each encounter, they said the same thing, a sad, ever-recurring refrain. Never was I allowed to forget my fatness. And with every year, there was more fat for them to worry about.

Saskia Clemency is born

Saskia Clemency was born in 1989. I was twenty-seven and would have been supremely content to stay at home with both my girls and make pasta collages. My babies fascinated me. I prepared Allegra carefully for the new baby's arrival, determined she wouldn't feel replaced and redundant as I had when they brought my sister home. I didn't let the Yiddish chorus loose on her. I wouldn't allow them to spout bilge about being 'a big girl now' and her 'needing to look after the baby'. She was only three. She didn't have to be big or look after anyone. I wanted her to have the chance to come to love her sister in her own time and at her own pace.

Left to blossom, the love between my girls was organic and unforced. Saskia idolized her sister. Allegra found Saskia hilarious. NB Saskia *is* hilarious. The two evolved their own relationship without any bossy grown-ups telling them what they ought to feel. Saskia was gregarious and intrepid. Quickly she became the duo's social representative, introducing more reticent Allegra to groups of unknown children with considerable swagger. She was a climber, prone to shinning up walls and doors, an entertainer, and enormously empathetic even as a baby. She has always been a harbinger of happiness.

13

Fat, fatter, fattest

APART FROM A FEW MOTHER-GENERATED flirtations with amphetamines, my weight climbed till I was around a size sixteen–eighteen. I wasn't grossly fat, but I had certainly begun to fulfil their doom-laden predictions. Thus began my Battle of the Bulge.

Speed and motherhood didn't mix so I dieted. I went on the special fourteen-day diet – and all I lost was two weeks. I tried the Banana Diet and put on half a stone. Don't be fooled by appearances: a banana is a cream cake masquerading as a fruit. My house was bursting with diet books: Hip and Thigh, Bum and Tit, F and Blind, Start and Fart. You could poach them all lightly in a little lemon juice and you'd end up with the definitive diet: If It Tastes Halfway Decent, Spit It Out! Every woman I knew was dieting. Daily, over snacks of dust and air, we debated the merits of the Beverly Hills diet, the F-Plan diet, the No-Diet diet. I was always desperate to lose weight. I was now in charge of my own biscuit tin. I had never been hungrier.

We didn't have a bean, but I didn't want to eat beans anyway. I wanted a few spoonfuls of ice cream, a KitKat, a couple of Bourbons or a Milky Way. Alcohol didn't do it for me. I couldn't dabble with drugs or have sex with strangers, so food became my illicit vice. It had to be a secret because, as the weight crept on, I sure as hell couldn't eat in front of my parents and was embarrassed to *fress* (tuck in) in public. I never demolished an especially large portion at the table but afterwards, under cover of clearing up, I'd swipe a couple of fingers of Fudge. If you'd followed me around, I don't think you'd have been disgusted. I wasn't the kind of eater who polishes off an entire leg of lamb at one sitting. I didn't throw up afterwards – though sometimes I wished I had.

The girls were little. Funds were non-existent. I was trying to make a living as a journalist, swell the family coffers, make chopped liver from scratch in the kitchen and be a whore in the bedroom. I was never altogether sure about that, though. I often thought I'd rather provide the conjugals free and charge for cooking. I was doing my utmost to be a dutiful daughter, reliable friend, charitable citizen and all-round decent human being. How terrible could it be if I ate a couple of slices of pie? Why shouldn't I have it when I needed and wanted it so badly? I wasn't a binge eater. I was simply 'evening off' the cake. You know, you cut off a tiny piece and you've made the sides uneven. So, you cut off another minuscule sliver and eat it. But the cake's not symmetrical. So, you do it again and again in miniature increments until there isn't much left and you might as well finish it off.

Our marriage, I thought, motored merrily along, partly because we were a dynamic team, doing our utmost to be upwardly mobile. Our young married mates were conquering corporate life. The accountants were upgrading to become financial directors and earning up a storm. The lawyers were becoming partners. The IDBs – in Daddy's business – were acting MDs and being handsomely remunerated. We were trailing behind. If you're a doctor in the NHS, it doesn't matter how much initiative, drive or talent you have, you must still plod through the system, putting in year after year at different junior levels before you earn the right to progress to consultant. Even a certified genius doesn't vault the process. The Good Doctor was earning so little, secretly I used to wish he was a kosher butcher. At least if he were putting chicken livers into tiny plastic bags, we'd have been able to afford the occasional dinner out and a punnet of strawberries. I was earning even less. Existing on an overdraft is stressful, limiting and boring but we had our love to keep us warm.

14

Working nine to five

I DIDN'T WASTE ANY TIME wanting to be famous. There was no point. People like me were never going to be snapped, like Lulu or Dusty Springfield, 'alighting from an overnight flight from LA' in a Mary Quant lamb-chop frill skirt and wet-look boots. We weren't alighting from anything at all. If you'd asked the young me if I would like to waft about in marabou-trimmed chiffon on television, breathing the same rarefied air as Larry Grayson and Lena Zavaroni, I'd have accepted in a trice, but I knew I was never going to fend off frenzied autograph hunters at the stage door of the London Palladium. I wasn't a stage-school kid hoofing away at Aida Foster singing 'On The Good Ship Lollipop'. I was wrestling with Latin verbs and learning chunks of *Paradise Lost* by heart. You don't get famous doing that. Cliff Richard lived just round the corner in Northcliffe Drive, Frankie Vaughan down the hill and lesser-spotted Beverley Sisters could be glimpsed at Fred Blackwell the greengrocer. They were glittering adornments insulated in an impenetrable bubble wrap of fame. You could watch them, fascinated, from afar. You could never, under any circumstances, become one of them. I'm a pragmatist. I only fantasized about the attainable.

Brief foray into porn

I was banging out articles on hair conditioner and beauticians' workwear for *Hair Flair* and *Hair and Beauty*. No one gets famous for writing about the season's must-have hairbrush, but I'd married a junior hospital doctor on £6,750 a year and the mortgage wouldn't pay itself. I turned my hand to writing anything that might make me a few quid. Yes, I did become the columnist on Paul Raymond's naked-lady mag, *Men Only*. Yes, I did write under the soubriquet Vanessa Focks. Yes, there were a few regular customers who would take delivery, then post the mag back to the Soho office in a sticky condition. I quite enjoyed concocting kinky tales from the post-lacrosse-match ladies' showers. It made a change from comparing six sets of heated rollers.

The organ of British Jewry

In a delightful juxtaposition of jism and Judaism, Pulitzer Prize-winning editor Ned Temko made me the first female columnist at the *Jewish Chronicle*, self-styled 'organ of British Jewry'. It was a controversial appointment. The page layout placed me just below venerated writer Chaim Bermant. He looked like Isaac Bashevis Singer, wild rabbinical beard billowing in the breeze, and he spoke Yiddish in a thick Scottish accent, or possibly impenetrably Yiddish Scottish. He was rightly revered. The soil of the *shtetl* still coated his boots, though he paced the streets of Hampstead Garden Suburb. He looked at least five hundred years old. I was twenty-seven. I was blonde and busty. Cue the hue and cry. One letter of many published the week following my inaugural column began: 'Vanessa Feltz is not fit to lick Chaim Bermant's boots.' I told you they needed licking.

People in Grodzinski's (kosher bakery) recognized me from

my *Jewish Chronicle* picture. I was accosted at Sainsbury's in Golders Green. My feature on au pairs – did Jewish families exploit these fair Eastern European virgins by asking them to spring-clean their houses to eliminate every particle of *chametz* (flour) before Passover? – caused havoc and the threat of a lawsuit. My undercover exclusive on three separate dinner parties – student, young marrieds, crusty socialists – divided opinion in Hendon Central. I was asked to speak at a fundraiser in Whetstone. I was asked to speak at a fundraiser in Brighton. I was asked to speak at a fundraiser in Cockfosters. I couldn't refuse. How can you tell Marlene Goldberg of Ilford you're too terrified to 'say a few words' to refurbish the rabbi's vestment cupboard?

I had to pitch up. I had to say something. But what could I talk about? All I'd done was snog on the Haberdashers' coach, go to Cambridge, get married, make chicken soup and give birth to two exquisite geniuses – the same, bar the Cambridge bit, as every single woman in the room. I was much too scared to extemporize. I couldn't trust myself to think of a single spontaneous syllable. So, I wrote down 11,979 words. I remember. I counted them. I stood up at the charity brunch/salt beef and latke buffet/ *lokshen* pudding tombola and read what I'd written.

Mum, Dad and Grandma Babs came to my debut. I galloped through the pages, too nervous to look up in case I lost my place. At the end, Vera Margolies asked for questions from the floor. My father, Norman Feltz, rose to his feet. 'Ms Feltz, may I say every word you have uttered is a pearl, every sentence a symphony, every paragraph a diadem. Would you say these are inherited characteristics?'

15

There is no parapet moment

How could I possibly have known the article I bashed out between attending to Saskia's nappy rash and defrosting six turkey thighs would propel me to fame – the British household name, tabloid fodder, camera-in-your-face-at-your-mother's-funeral kind you can never shake off, however hard you try?

What if the Lord had appeared to me in 1989 and said: 'Halt, Vanessa!'? (In fact, the Almighty would probably have used my Hebrew name. Did you know Jews have a Hebrew name in addition to their English one? If you think I'm telling you mine, think again. I even lied about it at Hebrew classes with disastrous consequences. FYI it isn't Shoshana.

Parents' evening, Woodside Park Cheder 1969. Miss Hackenbrough; 'You'll be glad to hear Shoshana is making wonderful progress.

My father: 'We're delighted for Shoshana's parents. How is my daughter doing?')

If the Holy One, blessed be He, had swooped down and issued a divine warning: 'Vanessa Feltz if you write that article, it will make you famous, infamous, phone-hacked, quite a lot of money, a beacon for blokes on the make, one of the celebrities asking the questions in *An Audience With* and unable, for the rest of your life, to put on a few pounds, have a bit of Botox or fall out with your father without the *Daily Mail* writing a double-page spread on it. Do you still want to write it?'

I think I'd have asked Him to wait a couple of minutes while I had a cup of coffee and pondered the implications. Alas, He didn't bother. He didn't issue a red alert. I had no idea that just plonking out another article, yet another article, to add to the thousands I'd churned out already would do anything more than earn me eighty-five pounds before tax.

I didn't choose to stick my head above a parapet

All this stuff about famous people 'putting their heads above the parapet, so they deserve whatever foul calumny they get' is bollocks. Sure, if you enter a TV talent contest, go on a reality show or try your utmost to be an influencer, you are actively seeking fame. You probably deserve whatever befalls you. The rest of us got famous by mistake, just by drudging along at our day jobs. We never had the chance to think, Woah, there! What's that? OK, that's the parapet! Right, I am now of sound mind, have weighed up the consequences and am taking the deliberate decision to crane my head and peer over the thing. It's not like that at all. THERE IS NO PARAPET MOMENT.

One minute even your auntie Carole struggles to remember your name. The next you're being paid, as I was in 1996, to say: 'Coming up, two great Birds Eye-deas!' I didn't mean to do it. I'm not saying I wouldn't have gone for it if I had known. I am saying I didn't, and I didn't. Fame! Forsooth! Being (in)famous has been a bumpy, brilliant, bellicose and brutal ride, but it's an accident I never signed up to.

First time on the radio

The fateful/fatal article was, I admit, provocative. I took a pop at the Jewish Mother. I had one. I was one. I was the perfect person to assert, in the *Jewish Chronicle*, that the sainted 'Yiddishe Momme' was extinct. The Jewish mum of 1991 was more likely to be at the gym, clutching a latte, enjoying a spell of 'me-time', leaving her precious children in the care of the au pair. I dimly recall penning the sentence: 'There are Jewish children under five speaking only Portuguese.' The paper published.

Outrage erupted. To say the *Chronicle* building was pelted with matza balls by a rabble calling for the editor's head on a pike would be a lie. The protests, however, were tumultuous. Affronted readers claimed their own selfless matriarch would keep her window closed at night so she wouldn't inhale a single puff of fresh air that could be breathed instead by her adored children. Pulitzer or not, the editor had no choice but to print a retraction the following week. I was invited on to BBC GLR 94.9's newly created Thursday-night show *Jewish London* to defend myself.

I'd never been in a radio studio before. Seated in the green room with a unicycling rabbi, the winner of the Finchley Synagogue challah bake-off and a Holocaust survivor, I was awash with terror and adrenaline. I sipped water to stop my throat silting up. My mind iced over. Internally I was an igloo. They shoved me behind the mic. The green light pinged on. Presenters Brigit Grant and Richard Brecker began their genial interrogation and something unexpected happened. I said a load of funny stuff. I was funnier than I'd ever heard myself be. Things I hadn't known I thought poured into the microphone. They were laughing. I was laughing. The rabbi and the Holocaust survivor were chortling.

I didn't know it then, but I do now. Most people are less funny/fluent/convincing live on air than in the bathroom at home. Watch politicians mumble, stammer and sweat as they struggle for a 'relatable' reply. Trying to speak naturally in a studio with professional interviewers doing their utmost to trip you up is difficult. The best raconteur in the green room turns into a disappointing flop in the studio. In the heat of the moment his punchlines evaporate. He loses his thread. He develops a weird verbal tic. He staggers, blinking, from the studio, horrified at what he said and tortured by what he didn't. Stand in the corridor and you'll see broadcasting casualties bent double with sudden-onset migraines, clutching their stomachs and gasping for breath. They couldn't say what they meant. They didn't mean what they said. They were mashed potato. They vow never to go near a studio again.

Not me. I was elated, inflated, aerated, so astounded by my own performance, I charged straight to my parents' new house in Highgate. 'Mum, Dad, did you hear me? What did you think?'

My mother: 'I've told you before. You must lose weight.'

Me: 'Lose weight? I've been on the radio.' Collapse of stout – she was right: I was too fat – party.

Objections and obstacles

Did I mention I didn't get paid to go on *Jewish London*? Did I mention my parents asking how my husband felt about my driving twenty minutes away from NW2 and 'footling about' on the radio when a married woman should really be at home attending to her husband the doctor's every whim? They pointed out that it cost me money to be a guest on the show. Remember how Norman the Knicker King raised me? I must factor in turning on the immersion heater for a hot shower, use of the blow-dryer, petrol, depreciation on Poppa Willie's ancestral handed-down Honda tyres, price of the babysitter – of course the Good Doctor could not be expected to step into the breach and tend his own offspring, not for a whole hour and a half – and a celebratory Twix to still my beating heart on the way home. I was quids out. I was skating on thin ice. My own parents outed me as too fat for radio. My husband couldn't see the point of my schlepping up to town to talk nonsense about synagogue fashion for the fuller figure.

You have no idea how shocking what I'm about to write is. I can't quite believe I'm writing it.

I DIDN'T CARE!!!!!!!!

For the first time in my life, I was drowning out the Yiddish chorus. Let them tut and kvetch. I didn't give a damn. It was hardly an earth-shattering gig – just a few minutes, unpaid, on exceedingly niche local radio, but it was a riot. I was a riot. I kept spooling the memory through my brain. I couldn't stop thinking about it. For once I badly wanted something that had nothing at all to do with marriage, children or trying to persuade my parents I wasn't a big fat anticlimax.

I wanted to be asked back and be on the radio again. They asked. I went. The green light glowed and so did I. They could fling a subject at me, any topic, matza, sex through a hole in a sheet, a debate about why Jewish people pave over their front gardens, and I riffed, spun vignettes and sparkled. Driving to Sainsbury's I'd count my appearances. 'I've been on the radio eight times. I've been on the radio twenty-seven times.'

Gloria Abramoff, backbone of the BBC, freedom fighter, ethnic broadcasting pioneer, fearsome foe, ferocious friend, didn't much like the cut of my jib. She thought I'd floated in from suburbia on a raft of privilege. She's a radio maestro. She listens to the melody and flow of words as if to a cadenza. She is confident enough to change her mind. She asked me to report on some crucial issue I can't recall. She paid me eleven pounds fifty. It was my first professional broadcasting fee. Eighteen months later, I was presenting the show.

16

Fergie's toe sucking

FERGIE THE DUCHESS OF YORK'S toes had been sucked by her 'financial adviser' John Bryan in front of Princesses Beatrice and Eugenie in a sun-kissed location, which meant scanty swimwear for her and budgie-smugglers for him. Photos of the salacious interlude, hidden behind the arras in some London apartment, had been unearthed by builders and sold to the highest tabloid bidder. The nation went chicken oriental, off our heads doolally. Toe sucking! We'd never dreamed of such debauchery. We were aroused. We were horrified. We were turned on to the max and so repelled we had to have another lengthy gander at the incriminating photographs.

Poor Fergie was still married to that glorious specimen Prince Andrew and they were having a sleepover at the in-laws. Imagine it. She had to descend to breakfast with a stony-faced Duke of Edinburgh, knowing he'd digested images of her pedicure disappearing down a Texan throat over his devilled kidneys.

What are the Duchess of York's sexual shenanigans doing in a book about me? I'll tell you. By 1992 I had inched a little higher up the greasy pole of magazine journalism and was writing columns for National Magazines' *She*. It was for *Cosmo* Girl grown up. It was sexy fashion and fashionable sex but made room for a few token features on nursery decor and demand feeding. Writing for *She* evoked hitherto unknown sensations. I enjoyed reading my own copy. It was a mag I'd have shelled out my own money (if I'd had any) to buy. I was the target market. I loved writing for it, partly because the ex-*Cosmopolitan* editor Linda Kelsey was on the ball and innovative.

Linda had been thinking about the toe sucking. Who hadn't? With customary elan, she decided to make it work for *She*. 'Vanessa, I've been thinking. If toe sucking is such a sizzling sex sensation, I want you to write a piece about the other obscure bits of the body that pack an undiscovered erotic punch. Over to you.'

I did a lightning survey on the fireworks-inducing parts of our anatomy languishing unacknowledged. I asked a few mums at playgroup and the dental hygienist and produced a sprightly feature highlighting the erogenous potential of the backs of the knees, navel, crook of the elbow and, most memorably, the top of a bald man's pate. Drawing heavily on the sultry and suggestive Häagen-Dazs ads of the nineties, which engendered in my generation an unshakeable belief that orgasms and choc ices were inextricably linked (confession: I believe it implicitly to this day – brand loyalty rules), I wrote, drawing largely on my own imagination, that a tablespoon of ice cream, gently massaged into a slaphead's dome and delicately nibbled off again, ignites volleys of curtain-trembling multiple orgasms. Try it for yourself if you dare.

She hit the newsstands. The reception was dementedly stratospheric. The *Sun* published a double-page spread on my findings. Its title: 'DON'T BE A THICKIE. GIVE A BALDIE A LICKY.'

It should have been the making of me. The landline, the other sort hadn't been invented yet, should have been ringing off the hook. Yet there was nothing. Zilch. *Nada. Rien. Gornischt*. Until ages later the wretched thing finally rang and it wasn't the

Golders Green branch of WeightWatchers. 'Hello. Is this Vanessa Feltz? Please say it is. Please.'

'Yes. This is she. May I help you?'

'Oh, my God! It's you. At last. Fucking Ada. I can't bloody believe it. You can't imagine what a flipping nightmare it's been to track you down.'

'What do you mean? Who is this? Why are you trying to get hold of me?'

'I'm ringing from *This Morning*. You know, with Richard and Judy. We've wanted you on for months. Ages. We kept asking your magazine editors to put us in touch. They wouldn't hear of it. They all said you were exclusive to their magazine and not permitted to go on TV. I can't tell you how relieved I am to hear your voice.'

Days later I boarded a plane at Heathrow on my way to Liverpool's Albert Dock.

17

This Morning *with Richard and Judy*

I WAS BOBBING BETWEEN GLR and *This Morning*, still unknown beyond Finchley Central. Once, in Sam Stoller's kosher fish shop, a woman pulled a bunch of paper-clipped cuttings out of her handbag and assured me she posted my column to her daughter in New Jersey every Friday without fail. I was elated until I remembered my column ran only once a month.

I was still earning nothing at all at the BBC and too little at ITV to make a dent in the all-engulfing, ever-expanding overdraft. Clothes cost a fortune in 1992. Primark didn't hit Oxford Street till 2007. I needed televisual outfits but didn't have the dosh to buy them. These days, you'd pick up something cheap and snazzy online. That wasn't an option. I still cringe at the memory of sitting opposite Richard and Judy decked out in heavily embellished suits borrowed from my mum. It didn't help that she was at least two sizes larger and – it's OK to say this, she's been dead for twenty-eight years – had the fashion sense of Fanny Cradock. My mother was many things, an LSE graduate in history, an inspired dinner-party cook – she was bored by ordinary daily dinners but whipped up pneumatic soufflés and rather often an ingenious seventies dessert known inaccurately as *bombe surprise* – but Coco Chanel she most definitely wasn't. Imagine the embarrassment.

Just thinking about it now, my cheeks are turning purple. Millions of innocent people forced to look at me in their living rooms, wearing a baggy Crimplene 'leisure' two-piece in apricot with diamanté appliqué.

Looking back, I really should have indulged in a little light prostitution or at the very least driven a minicab in the evenings to raise funds to kit myself out in less offensive attire. Why didn't I flog my gold fillings or pawn my engagement ring to save myself and the audience from another Valerie Feltz ruffled neckline and pleated skirt in acid green? Why didn't someone who was meant to love me – husband, parents, just saying – chip in a few quid to rescue me from Frump City? These days I have my own fashion collection, Vanessa4love, and Instagram followers say they never see me in the same dress twice. Now you know why. Other people buy art and shares in pork-belly futures. I buy dresses. The horrible feeling of *literally* having nothing to wear – not nothing you fancy wearing, I'm talking about nothing except the dress you wore last time and the time before and the time before that – has never left me. For me, even now, too many dresses are never enough.

Life was hectic – see chapter on guilt. Whoops, I haven't written a chapter on guilt but I do feel guilty about it. The girls were six and three. A remarkable thing happened. For the first time, I looked forward to doing my job. I wasn't just chaining myself to the desk and cranking out articles as if I was doing my school geography homework. Work was suddenly social. It was unpredictable. It was, just a little bit, showbiz-ish. It was beginning to verge on just a fraction glamorous. Richard Madeley in a Mr Blobby outfit? Judy Finnigan taste-testing freshly frothed piña coladas? Sitting on the pontoon in the Albert Dock canal in bright sunshine discussing Andre Agassi's hairy chest? What's not to love? Throbbing with constellations of film, pop and TV stars, *This Morning* was the ultimate – 'Oh, my God, that's Ronnie Corbett! Jesus, isn't that David Essex coming out of the gents?' – working environment.

I'd spent seven years staring at the word processor and doing twenty-two creative things with mince. I pretended schlepping all the way up north to hobnob with Richard (catalogue-chap handsome) and Judy (pretty, petite and not suffering fools, especially Richard) was exhausting. It was bloody marvellous.

18

Whatever you do, don't admit you're enjoying yourself

I DIDN'T DARE SAY I was loving every minute because, let me be clear, there was a fatwah on fun in 1992. Married women with children were emphatically not permitted to enjoy themselves unless immersed, J-cloth in pocket, hand up a chicken's rear pulling out the little plastic bag of entrails, in the bosom of family life. Work was allowed but purely as a necessary evil to be endured. You weren't allowed to whoop it up on the BA flight back from Manchester with Paul Weller or rub shoulders with Basil Brush. I tried not to show life had started to twinkle. Pleasure? Excitement? Who – me? Don't be ridiculous.

I was so riddled with guilt I couldn't allow myself to savour a single sparkly second. Even when I was allowed to sail up to Liverpool with my arty, crafty cousin Beverley-on-both-sides. Her mum Auntie Marilyn and my mum Valerie were first cousins who married first cousins Uncle Harold and my father Norman. H & M fixed up N & V. My mother had standards: 'Norman must phone me and ask me out himself.'

My father rang and said: 'They tell me I have rather nice eyes.' He was right. Valerie lost herself in them, then V & N managed to lose M & H on purpose at the Earl's Court Motor Show. They gave birth to us. Bev and I are cousins on both sides, front side and back side. The two families intermingling meant when we were bridesmaids – a role we played so professionally we notched up at least seven stellar performances in Sonia Lyman dresses, charm bracelets and white satin ballet shoes – her grandma Ida and my grandma Babs, sisters, neighbours and pals ably aided and abetted by her grandma Jean and my poppa Willie, brother and sister, neighbours and pals, would be in the front row of the synagogue, cheering us on. Being cousins on both sides is a very special inextricably conjoined kind of thing. Don't knock it till you've tried it.

This Morning asked if I could fly up and make an 'alternative guy' for fireworks night. I couldn't. I'm a Chaucer aficionado, not a visionary kapok-stuffer. Cousin Bev, on the other hand, an internationally acclaimed window-dresser with an A-list roster of clients, is a dab hand with a pair of tights and a staple gun. I transported her Richard and Judy-wards. She killed it, miraculously creating, out of wisps of fabric and pipe cleaners, a feisty feminist female guy with attitude. Naturally, unleashed on the frozen north, we cousins had a blast to end all blasts and felt so dreadful about laughing ourselves stupid while our starving orphan children – hers are Ben, Sam and Joel – withered and perished from cruel neglect, we even pretended to each other we'd had a day of arduous wretchedness.

19

Check your husband's testicles for lumps. Here's Vanessa Feltz

THE INVITATIONS TO SPEAK AT charity lunches and dinners swelled from a tentative trickle to a Noah-esque flood. Naturally, the gigs were gratis. You can't charge a charity. Not even for petrol and a new pair of tights. Not even if you're married to an NHS junior hospital doctor and the mortgage rate has risen to 14 per cent. What's more, I knew I had to say, 'Yes!' to all requests – our fates depended upon it. Say 'No!' and I'd ignite the wrath of, you know who, the Evil Eye, ever-present threat to Jewish happiness.

Here's how the Evil Eye operates

If you are reckless enough to appear in public in new boots and a diamond tennis bracelet looking pretty, healthy, happy or wealthy your relatives will rush to protect you from the EE's jealous wrath. 'What lovely boots, *k'neyna hara*!' Translation: 'What? Are you crazy flaunting that flashy footwear? May the Evil Eye not wrench out your tongue for strutting around in such showy Louboutins!' If something or someone is just too ebulliently gorgeous not to attract the EE's attention – a bouncing baby or a top-notch blow-dry perhaps – concerned relatives will raise their first and second fingers towards their lips and make a spitting sound, *pa pa pa*. No saliva is sprayed unless they get carried away. This is notional spittle. The pretend spitting noise is supposed to ward off the cursed revenge of the dreaded EE. It seemed rational to me that if I didn't turn up to Catford Synagogue's 'Say No to Diabetes' whist drive my entire family would have to start injecting insulin the following morning. Refuse the rheumatoid arthritis people and you've only yourself to blame when all your relatives are – Heaven forbid – stricken with the disease. I never risked it.

I spoke at so many fundraising functions, very gradually I started to get the hang of it. My timing tightened. I took the plunge and uttered the occasional extemporized syllable. Once my heart rate calmed, I noticed the spur-of-the-moment stuff went over bigger than the material I'd laboured over in advance. The studio green-light effect was happening in Gillian Steinberg's rag-rolled turquoise reception room in Kenton. I had no idea what I was going to say until I said it. On the way home I couldn't remember a word. Yet the assembled throng laughed till their girdles ached.

TV gives you a warm-up. Charity functions give you Misery Man

Please note, fundraisers inevitably involve a gut-wrenching speech from the person I have no choice but to call Misery Man. This bloke is the opposite of Warm-up Guy who perks TV studio audiences into a receptive, light-hearted mood before the show with whistle-stop one-liners and prizes for the best rendition of 'Agadoo'. At charity functions, Misery Man from the committee takes to his feet and hectors the audience on the deadly disease bringing victims out in noxious suppurating sores. He often uses illustrative slides. He's just doing his job – to make everyone feel so panic-stricken they reach into their fake Fendi baguettes (snapped up for a song on the beach in Marbella) and make out a substantial cheque to ensure the Lord doesn't smite those they love with boils. Misery Man warns: 'Check your husbands' testicles for suspicious lumps!'; 'The synagogue roof has turned to cottage cheese'; 'All those holidays in Marbella have probably given you melanoma.' Everyone feels awful. They write a cheque so large they wouldn't dare tell their husbands. Then they feel even worse.

At precisely that moment, Misery Man subsides on to the sofa and the hostess announces: 'And now Vanessa Feltz, who isn't fit to lick Chaim Bermant's boots.' These ladies had survived the world's most depressing diatribe, seen pictures of swollen spleens, eaten so much chopped liver their poppers were popping, and I still managed to make them laugh. Eventually, I dared to throw my script away and trust myself to dredge up something funny to say on the spot and that's when the late Peter Moss approached me.

Lord knows why Peter was hovering about the sidelines at a ladies' lunch. Perhaps he wandered in to collect his wife. He bowled over, fizzing with bonhomie, and said: '*Mazeltov!* You do know you're very funny, don't you?'

No one had ever put it quite that way before. My father was funny. Legend had it my late Clark Gable-clone grandfather had them rolling in the aisles. Me? Funny? I had just about begun to hope so, but there'd certainly been no official confirmation. Me: 'Thank you. It was a tough crowd after all those close-ups of necrotizing kidneys.'

Peter: 'I noticed. Have you ever considered doing stand-up?'

20

Have you ever considered stand-up?

CONSIDERED DOING STAND-UP? WAS THE fellow *meshuggah* (insane)? Considered doing stand-up? He might as well have asked if I'd pondered a career as a Red Arrow or tattooist. Stand-up? Not only had I never considered it, I'd barely seen it since university and hardly bothered with it even there.

I'm not sure what happened to me. I heard my own voice replying: 'Yes, of course. I've always longed to do stand-up. It's been my dream.'

My pants should have burst into flames. Pinocchio Feltz. Calmly, as if it was something I did every day, no biggie, I agreed to join Peter in a stand-up comedy show at Hampstead's New End Theatre. The dates were set. My name was on the dotted line. The posters, 'VANESSA FELTZ IN GEVALT THERAPY', were being printed before I'd driven home to make meatballs for the girls' supper. I'd been booked to headline professional comedy gigs. Peter, who'd been playing the comedy circuit for years, would be my support act.

Luckily, we had a tiny en-suite. I locked myself inside and tried to keep breathing. It wasn't easy. I'd signed up to deliver an hour of rib-tickling comedy to a live audience on my own North-west London doorstep. I wasn't a comic. I couldn't stand up in heels without needing a good sit-down. And – oh, yes, significant obstacle alert – I didn't have any material. None. I'd never done actual jokes. I'd been incidentally amusing but until that moment if I wasn't funny it didn't matter. I could be fluent, peppy, opinionated. That sufficed. I hadn't claimed to be a comedian. I didn't think of myself as a comedian. I didn't do punchlines. I had no idea how to write a routine. I didn't have an inkling how I'd fill the first two minutes, let alone a whole hour. Also – hardy perennial problem – how would I manage to lose enough weight in time for my stage debut? I needed to shed a stone and a half and come up with sixty minutes of spine-tingling jokes in just four weeks. No wonder I polished off a whole packet of Bourbons.

The actor and polymath Dave Schneider once told me comedians don't sit down and write an hour's act all at once. Apparently they evolve their material in five-minute increments, spending months fine-tuning and audience-testing. They might start at an open-mic night with three or four minutes and work slowly up to a full fifteen-minute act five years later. The thing is, he didn't tell me that till much too late. At the crucial time I didn't know anything at all about the comedic process. All I knew was I had to think up enough funny stuff to stop the audience (if, indeed, anyone showed up) fleeing the theatre, and I had almost no time to do it.

That Sunday morning, I persuaded my husband to take the girls out so I could get stuck in. They were gone for about three hours. When they returned, hungry for lunch, I'd written a shtick about Princess Anne's husband Captain Tim Laurence's family secretly being Jewish. Rumour, I said, reverberated round Edgware that the Laurences' real name was Levy. I described Tim's mother, Mrs Laurence née Levy, hosting the *aufruf* and the Queen and the Duke of Edinburgh showing up in *shul*.

Don't tell me you're wondering what an *aufruf* is? You mean you don't know? An *aufruf* happens a week before a Jewish wedding at the groom's family's synagogue. The groom is honoured by being called up in front of the ark to read from the Torah. The groom's side grabs the chance to overshadow and outshine the wedding by hosting a luncheon in a marquee festooned with pictures of the man of the hour aged six months, naked on a fur rug. Guests enjoy a smoked-salmon station, a salt-beef station and a life-size sculpture of High Barnet station, where the bride and groom met, constructed entirely of *cholent* (a slow-simmering stew). *Aufrufs* are funny. Her Majesty as mother of the bride taking her seat in the ladies' gallery at Bushey Synagogue is funny. I hoped, somehow, I could whisk the audience on an imaginary excursion into an excruciatingly embarrassing *simcha* and make them laugh.

Comedy testers

Quickly I realized jokes need human testers. You can't just write them, say them aloud to yourself and work out if they'll fly or bomb. My testers were duds. I'd pepper-spray my husband with gags on recycling by weaving ripped-out superfluous hair into cosy *kippot* (skullcaps worn by Orthodox Jews). I'd deliver my lines with panache. He'd sit stony-faced. 'Is it funny?' I'd ask beseechingly.

'I don't know,' he'd reply, deadpan.

I tried my mother. Valerie Feltz wasn't big on humour, especially of the vulgar variety. I'd unleash a humdinger about sex in the rabbinical position and she'd insist I wash my mouth out with Imperial Leather. She was a one-woman censor, terrified her pals from Potters Bar Golf Club would buy tickets and be corrupted. No sooner had I written a joke than she'd vetoed it. She was a fatwah in female form.

I had to find someone else to experiment on. Enter our Polish au pair Monika. She wasn't exactly my target market. She was twenty, fresh off the plane and barely spoke a word of English, let alone Yiddish. Thank God for Monika. Instead of stripping the beds and helping with bath-time, I promoted her to full-time laughter monitor. OK, she didn't have a choice in the matter. She lived in my house. She couldn't escape me. She was the quintessence of a captive audience, but she obliged with jovial enthusiasm and, through some quirk of fate, I cracked her up. She may not have understood the nuances, but Monika appreciated a good dirty joke when she heard one. She was a fantastic sounding board. Fuck the language barrier. Most of the time she laughed her head off. If she didn't, I binned the joke. Simple as that.

By day, I feverishly hammered out material. By night my husband and I guerrilla-plastered posters advertising *Gevalt Therapy* all over Hampstead and its environs. We were a crack clandestine team. That's part of the reason I was so blindsided when he left eight years later. We had such larks jumping out of our Talbot Alpine with brush, pot of glue and rolled-up posters, slapping them up and scarpering. We were on the same side. I hoped I'd hack it as a stand-up. He hoped the same thing. We were in it together, and I knew if I crashed and burned, he'd apply the bandages.

21

Opening night

THE SHOW WAS A SELL-OUT. No one knew how or why. The only logical explanation must have been Peter's reputation, faithful followers from the *Jewish Chronicle* and a dearth of anything else worth seeing. Inspired by Dame Edna Everage, who pelted her audience with gladioli, I decided to throw fried gefilte fishballs at mine. 'There's an ancient Talmudic tradition,' I told them. 'If you eat a woman's fishballs, you must laugh at her jokes.' Strictly speaking, they weren't 'my' fishballs. I'd bought them from Sam Stoller's, kosher fishmonger of Temple Fortune, that morning.

Yes, they did impart a certain *haimishe* (homely) aroma to the New End Theatre. The place reeked of carp and onions. I hoped for a Proustian effect. One sniff of the familiar scent imbuing their living-room curtains with generations of pescatarian perfume, and my audience would feel nostalgic, relaxed. I hurled the balls. Under-arm. You've never seen Jews move so fast. The hunter-gatherers settled back in their seats, a collective chomping as mass mastication ensued amid an atmosphere of amiable anticipation.

They lapped up Princess Anne's *aufruf*. They howled at 'Do Jewish girls swallow? Is sperm kosher?' Grandma Babs, sitting in row B, was so sweet she managed not to hear the perfectly audible lines that should have shocked her rigid. The *Jewish Chronicle*'s glowing review called it: 'Immensely funny, explicit, more blue than white, this sell-out show isn't one for prudes.' (NB I still don't really understand 'more blue than white'. Is it some sort of pun on the colours of the Israeli flag?) It was official. I was a stand-up comic. I was chuffed to bits. There was just one problem. I had to do it all over again the next night.

The night after

Seasoned comics will know what's coming, but I wasn't seasoned. I wasn't even lightly salted. I hadn't the faintest idea of what was about to befall me. The next night I stood on the same stage in the same theatre at the same time, wearing the same heavily beaded jacket from Faimans of Brent Cross. I delivered the same jokes, with the same intonation. On Saturday night, there'd been peals of laughter. Joel Zimmowitz chuckled so hard he thought he'd given himself a hernia.

Sunday night, I found out, right there, solo on stage at the New End Theatre, is not Saturday night. Saturday audiences are frolicsome. They are looking forward to a lie-in and a post-theatre shag. They nibble on their beloved's earlobe mid-performance. They are Travolta in *Saturday Night Fever*. They're up for fun.

Sunday night is a different ball game. They've had it to the back teeth with their spouses. They're still full of lunch and wishing they'd refused seconds of strudel. Their thoughts turn to toil and *tsores* (worries). Making a kvetchy, tetchy Sunday-night audience laugh is hell. I rattled through my script, hoping some of my shtick might stick. It wasn't a Titanic failure (iceberg, Goldberg, what's the difference?) but it nearly burst my bubble of confidence. I had to learn on my feet with everyone watching me. Sunday-night audiences need TLC, to be gently cajoled out of their malaise. It takes more than a fishball to bring them on board. I tempered my tone and smiled more. I acknowledged at least half the bums on seats belonged to blokes who were only there because their wives made them come.

For the first time ever I wished I was famous

For the first time in my life, I wished I was famous. How bloody wonderful would it be, I thought, staring at a (Red) sea of stern faces set in 'Go on, then. Impress me!' mode, if people felt pleased just to bask in my presence. If I were Barbra Streisand, they'd be privileged to kiss the hem of my Armani. One bar of 'Evergreen' and they'd leap in the air, baying for an encore. If I were Maureen Lipman I'd say, 'You got an ology!' and they'd erupt like lottery winners. What a monumental relief it must be to be a celebrity. People think they've scored just by breathing the same air. If you sing, dance or make them laugh it's a bonus. If you don't, they still snap up the anti-dandruff shampoo you're endorsing. Half the battle's been won before you even set foot in the auditorium. More than half. They already love you. They want to enjoy you. They're admirers, predisposed to think you're great because that's why they schlepped here. I'd never wasted a second of my life wanting to be famous before. Suddenly I didn't just fancy fame: I understood what a terrific resource it could be. I wasn't averse to adulation either. Who doesn't adore being adored? Doing my utmost to thaw my deep-frozen Sunday-nighters, I said a quick, slightly self-conscious prayer. 'Go on, then, God, please make me famous. I could do with the PR.'

Why didn't the audience rise in unison and warn, 'Vanessa Jane Feltz [insert Hebrew name, but you can't because you don't know it], BE CAREFUL WHAT YOU PRAY FOR!'

22

Every Jew is a critic

EVERY JEW IS A CRITIC. If you don't believe me, give stand-up a whirl. I popped out to buy a chicken, and Greenspan the kosher butcher told me the shtick about kosher butchers needing trimming. Martin Hofman, in the family chemical business, said the swallowing bit should have included gags about gagging. Suzanne Mendel thought I spoke too fast. Arlene Mindel thought I gesticulated too wildly. Rebbetzin Mandel (the rabbi's wife) wondered if I should have incorporated a few Yiddish songs, just for variety. Uncle Clive thought Peter Moss was fantastic. I privately thought that was for Peter Moss's uncle to express, not mine. Lionel the hairdresser at Highlights a Go-Go in Stanmore felt the chicken-soup part went over big but wasn't so sure about Singapore 'so clean you can eat off the floor in the toilets'. He'd been there. It wasn't that clean. Simon Israel, North-west London's favourite travel agent, thought I should have mentioned Cunnilingus the Irish airline. Lloyd Levy, optician, worried about the outfit. 'Not so sure about the beaded jacket, *boobelah*. It's not doing you any favours.' My GP, Nolan Wengrowe, liked the 'going private to have a baby because you get a better class of child than on the NHS' bit, but not the line 'I've always blamed the *mohel* [circumcisor]'.

My parents repeated the familiar refrain. 'Why are you so fat?

In that jacket, they could see you from the Tel Aviv Hilton. Grapefruit. Melon. Melon. Grapefruit.' By the time they'd finished I was so ravenously starving, I had to stop off at Grodzinski's for a quarter of Florentines.

We take the show on the road

My old school pal Matthew Kalman, editor of subversive Jewish magazine *New Moon*, popped over. 'Vanessa, you've done the hundred-seater New End Theatre thing. It's time to think big. *New Moon* will go halves. We can't afford to mop up the whole bill. You're in for fifty per cent. Let's hire the Shaw Theatre in Euston . . . I know it's in Euston. I just said so, but technically it counts as the West End. Four hundred seats. We'll take it for four nights. We only need to flog sixteen hundred tickets.'

Sixteen hundred freaking tickets? I panicked. Hang on! Hold your flipping horses. Sixteen hundred? 'Matthew, be realistic. Think. There's a pitifully small pool of potential punters. Do the maths. Number of Jews in the United Kingdom – fewer than a quarter of a million. Take out of the equation the old, the infirm, babies, toddlers, kindergarteners, expectant mothers, exhausted *booba*s, breast-feeders, Mancunians, Glaswegians, South Londoners, Communists, hay-fever sufferers, anarchists and everyone else with something better to do, and get real. If every single solvent Jewish stand-up comedy aficionado in the country purchased a ticket, would the total add up to sixteen hundred?'

What's more, I wasn't Florenz Ziegfeld. I didn't have his money. I didn't have any money. We owed Mr Barclay, of Barclays Bank fame, a princely four-thousand-pound overdraft obtained on the basis that one day, if he slogged in the lower ranks of the NHS for long enough, my husband, the surgeon, would make enough in private practice to pay it off. I didn't fancy telling the Good Doctor

I wanted to blow two thousand on hiring a fancy-schmancy theatre to show off in. I couldn't mention we'd lose every penny if we didn't sell the seats. I knew I was neither Barnum nor Bailey. What the hell did I think I was doing, playing fast and loose with the overdraft? Channelling Metro, Goldwyn and Mayer?

I paid a quiet visit to our local bank manager. Banks had real flesh-and-blood people working in them in 1993 – life was so quaint in the olden days. I grasped a fresh loaf of Hovis bestowed upon me by that sweet youth in a peaked cap who used to pedal that pushbike up the cobbled hill in rhythm with Dvořák's 'New World Symphony' and begged. The bank manager caved.

Matthew and I pretended we knew how to be impresarios, recycled the *Jewish Chronicle*'s glowing 'more blue than white' verdict, printed it on posters and festooned them, under cover of darkness, in the darkest recesses of Southgate. We hired Jonah and the Wailers – an a cappella barbershop quartet with an Israeli frisson – featuring Dan 'creator of *Whose Line Is It Anyway?/ Mock the Week*' Patterson and his chums – plus Dave 'Go on. Pick a matza. Any matza' Schneider and Clive Panto (his real name) as MC. Who did we think we were? I'll tell you: penniless speculators pretending to be Cameron Mackintosh. I couldn't afford sixteen hundred of Sam Stoller's gefilte fishballs. I bought a black net dress with a swishing pink crinoline skirt.

People kept telling me they'd been at the New End, loved it – apart from the foreplay gag, which didn't work – and they'd already booked the Shaw. I was appalled. Bloody hell! Repeat business meant more material. I couldn't recycle 'Do Jewish girls swallow?' if they already knew the punchline. *Oy gevalt!* What to write about? I decided on the local vocal yokel critics, the surveyors, gynaecologists and diamond dealers, who assassinated my performance with such savage assurance you'd have thought they'd played the Borscht Belt alongside Milton Berle. I named, shamed and defamed. Were they insulted, affronted? Are you kidding? They loved it.

Matthew and I didn't lose our shirts, or our *gatkes* (long knitted underwear). We had four nights of full houses. We were galvanized. Hiring sherpas, a St Bernard and some crampons, we schlepped the whole production up to Manchester. Yes, anthropologists confirm there's a Jewish tribe in the frozen north and they are Jews, but not as we know them. Down south we read the *Jewish Chronicle*. Up north they read the *Jewish Telegraph*. Down south we relish the sea-bass ceviche of Mr Tony Page. Up north they fine-dined on Celia Clyne. The language barrier was impenetrable. To stand a chance, I'd have to translate the whole show into Mancunian.

We rocked up the M1. Our tickets sold like hot *kugel* (noodle casserole). Luckily, the *This Morning* job was shaping up most unexpectedly, furnishing me with a whole lot of fabulous fodder.

23

Leonardo DiCaprio, Omar Sharif and Johnny Depp

Hollywood stars didn't always fancy making the odyssey to appear on *This Morning* in Liverpool. London was exotic enough. If they refused to budge from the Presidential Suite at the Dorchester, the programme had to send someone to interview them in W1. In 1993 sometimes that person was me.

Leonardo DiCaprio starred in his break-out movie, *What's Eating Gilbert Grape.* Sitting through a screening of the movie was the first hurdle. What a misery-fest! I loathed it, but Leo lit up the screen. I trotted to the hotel to find a monosyllabic nineteen-year-old wearing a woollen beanie hat and surly slouch. Forget Hollywood superstar, Leo was a typical truculent teenager. I couldn't tell if he was shy, hung-over or pissed off. Clearly, he'd never heard of Richard and Judy and would rather have had his balls stapled together than bother being charming to a no-mark like me.

I tried to prod him into sprightly conversation. He kept most of his face buried in his collar and mumbled like Marlon Brando in *On the Waterfront*. I was hardly TV's most experienced showbiz interviewer. OK. I was TV's never-done-this-before, fuck-I'm-out-of-my-depth novice, and he wasn't helping. Desperate to jolt him out of his uncooperative funk, I came up with an emergency strategy. Ask something unexpected. Shock him into submission. 'If you weren't stuck in this room with me right now, what would you rather be doing? Who with? And in what position?' He pushed his beanie out of his eyes, sat up straight and took notice. I can't remember anything he said, but he perked up, engaged and *This Morning*'s bosses were mightily relieved.

Forcing Leo to cough up some conversation was just the start. Backed by an intrepid *This Morning* crew, I accosted Sylvester Stallone and was rewarded with a deep, fabulous French kiss from Rocky in Leicester Square in full view of his fans. I know this would be an unforgivable #MeToo today. In 1993, I'd been married for eight years, and a bit of sprightly tongue action from one of the world's most charismatic chaps – though surprisingly short, I'm five feet two and a quarter and our embrace was eye to eye – was surprising but (understatement) not entirely unwelcome.

I interviewed my idol, Omar Sharif. I'd worshipped him since my seventh-birthday outing to the Finchley Odeon to watch him as Nicky Arnstein seduce Barbra Streisand's Fanny Brice in *Funny Girl* and had to exercise every scrap of self-control not to wriggle and giggle in his fragrantly after-shaved presence like a starstruck bobbysoxer. I grilled Johnny Depp and Belinda Carlisle. I was dispatched to Disney World in Florida to gush as the first ever British couple tied the knot in one of Walt's Beauty and the Beast weddings. Talk about impostor syndrome! I was acting like a seasoned TV interviewer/reporter/presenter, walking the walk and talking the talk, hoping if I projected calm confidence no one would realize I was winging it.

Other shows started booking me, among them John Stapleton's morning 'heated debate', *The Time, The Place*. John and his exhausted team trailed from Land's End to John o'Groats so *The*

Place would be different every show. Other presenters slept in their own beds and commuted daily to the same studio. John was permanently on the road just so he could announce: 'The Place? Coventry! The Place? Swindon!' In these cost-conscious times some savvy pound-stretcher would say: 'Is there really any value carting this show round the country like a travelling fair? Change the bloody title and give everyone a break.'

Producers briefed me to come to the aid of flagging programmes. 'Vanessa, if there's a lull, chip in. We're relying on you.' John was on holiday. A female presenter who shall remain nameless – mainly because I've forgotten her name – stepped in. 'Vanessa, if you see her floundering, pitch in with stuff on the importance of foreplay. Don't let the chat stall.' I was flattered. I'd only been on telly for a matter of months. When the debate headed for the scuppers, I waded in with both wellies and revved up the convo. I loved it. It was just like the school debating society, and I was being paid – result.

24

New York calling. Maury Povich *wants you*

I wrote a piece for *She*: 'You can be fat and sexy!' Technically, I wasn't fat. I was a sixteen heading for an eighteen, edging towards amplitude. The column made waves – I'm not exaggerating – all the way across the Atlantic. I picked up the phone to a producer on the coast-to-coast US talk show *Maury Povich*. 'Vanessa? Maury is mad about your piece. Fat! Sexy! British! Oh! My! God! We want to fly you over. We want to base an entire show around you. We don't pay but it's club class and you'll stay in a five-star hotel near the studio.'

As my dear friend the late David Gest would say: 'Holy focaccia!'

We were still surviving on the indulgence of Barclays and an ever-expanding overdraft. Holidays, apart from one self-catering in Lanzarote, total cost three hundred pounds all in, when Allegra was three, were off the menu. I begged my parents to look after the girls and cashed in the club-class ticket for two seats in steerage for the Good Doctor and me. Start spreading the news. A thousand words I'd bashed out on the word processor on my dining-room table propelled me all the way to the USA. What would Maury expect from me?

I pitched up at a shabby downtown building to be greeted on the pavement by Maury's excitable production squad. 'Vanessa?

Darling! Wow! Are we glad to have you? All you need to do, doll, and this is simple, make love to Maury with your voice. Make love to Maury with your face. Just make love to Maury. But make it British. OK?'

OK. I was whisked into Make-up, plastered in orange foundation, hair sprayed into a solid helmet. Backstage I was given the thumbs up from Maury. He's eighty-five now, bless him, and was hosting his daily talk show till 2022. In 1993 he was only fifty-four, but in the ghoulish spotlight, he appeared ancient, pickled in formaldehyde and done up in pearly pink lipstick, sparkly lip gloss, eye shadow, eyeliner, mascara and rouge. Used only to our very own Richard Madeley and John Stapleton, who required no more adornment than a manly slick of Brylcreem and a dash of powder on their noses, I was mesmerized. Maury had on more slap than a chorus girl. I couldn't stop staring. And I was on. A mate who happened to be in the US at the time took a picture of the TV screen. There I am, in my *yom tov* outfit, straining to look sultry. The strap line reads: 'VANESSA FELTZ – THINKS FAT IS SEXY'.

I must have delivered for Maury. They asked me back twice. Forget club class. I was promoted to first, my only experience of the American Airlines caviar buffet. I know. A buffet! Exclusively of caviar! Talk about life in the fast lane. Handmade chocolates in shades of lime, neon orange and Schiaparelli pink awaited on my pillow. I still wasn't paid, but thank you, Lord, for a dip in delightful, delicious, de-lovely luxury. We still couldn't afford a weekend in Brighton, and this was my third trip to America in a few months. Show number three: 'I'm Wild for Guys in Uniform' was the business. The finale – I wasn't forewarned or forearmed – involved being swept off my feet and carried out of the studio in a classic fireman's lift by a burly, six-foot-five, New York firefighter. Let's just say, I've had worse days at work.

25

What Are These Strawberries Doing on My Nipples? I Need Them for the Fruit Salad!

I HADN'T STARTED TO BELIEVE I had a career. The jumble of bookings didn't amount to one. None of the puzzle pieces made up a coherent jigsaw picture. Everything was piecemeal and unexpected. A bit of *Maury*. A bit of *This Morning*. Magazine columns. A touch of stand-up. Oh, yes, and I was commissioned to write a book. An enterprising literary agent called Judith Chilcote cold-called and asked if I had a book in me. I wanted to say: 'No, thank you. I can write anything in two hundred words or fewer. The idea of being chained to the word processor till I grind out an agonizing seventy-five thousand words brings me out in hives.' I didn't. I couldn't afford to. I needed the dosh.

I dreamed up what I hoped was a catchy title: *What Are These Strawberries Doing on My Nipples? I Need Them for the Fruit Salad!* It was billed as 'A Guide to Sub-Duvet Life As We Know It.'

The thrust – and it was all about thrusts – was long-term monogamous sex and how to survive it with sense of humour and relationship intact. Chapter headings included: 'If God had meant us to lick them, surely they'd have chocolate on the top?' and 'Sexual adventure – or how do you get taramasalata stains off the Laura Ashley?' Rather heartbreakingly in retrospect, I dedicated it to the husband I loved who – as you can clearly see – I thought loved me right back. 'For my darling husband – soul mate, sole mate, whole mate, hot date – with love and gratitude for your love and latitude'. Judith pitched it. There was, to my stunned amazement, a bidding war. She came away with a chunk of money and a two-book deal. There was just the tiny hurdle of writing the thing to be overcome. Nothing's more boring than banging on in one volume about what a fag it was wringing out another. Suffice to say I've happily allowed exactly thirty years to elapse between books.

Book two of the two-book deal never happened. I gave back the money. Here's why.

26

We want to make you the British Oprah

IN TELLY EVERYONE THINKS OF the same thing at the same time and tries to make it happen before a rival TV company gets their version up and running. In 1994 all the regional ITV channels were scrambling to produce a British version of Oprah Winfrey's confessional American talk show. The aim was to recruit the perfect presenter and persuade ITV head honcho Marcus Plantin to commission a series. Granada were lining up Bet Lynch, aka *Corrie* star Julie Goodyear. STV in Scotland had the bright and incisive Kaye Adams in their court. Anglia TV in Norwich fancied Feltz.

Malcolm Allsop, lay preacher, TV visionary, rang me. 'Vanessa Feltz, you are killing it on *The Time, The Place* [also an Anglia show]. We want you to be the new Oprah. We are going into partnership with Multimedia in New York. They make the Phil Donahue and Sally Jessy Raphael shows. The Americans have never heard of you. We made up a show reel of your appearances so far, but it's not enough to convince them. They've seen you talking, but not listening. They need to know you can listen. You must show compassion, empathy, all that kind of stuff. So, we need you to come to Norwich. Stapleton's doing a show with redheads. When he's finished, you'll step in and we'll film you with his audience, doing a hell of a lot of compassion, empathy and listening but, of course, also being sparky, funny, dynamic, all that malarkey. OK?'

My top lip stuck to my teeth. I watched in the wings as John Stapleton did his professional utmost to coax golden TV from a studio of gingers. His face was purple. A pulse throbbed in his forehead. The redheads were shtum, leaden, not giving an inch. John turned linguistic cartwheels. The gingers spoke rarely and only in monosyllables. I was hyperventilating. If John couldn't ignite them into some semblance of life, how was I going to show Multimedia top brass, awaiting the tape in New York, that I could not just listen but listen like Oprah? How could I turn these taxidermized gingers into entities who might *say* something I could pretend to the Americans I was empathetically listening to? If they didn't talk, how the beeping hell could I be filmed listening? John mopped his glistening brow with a handkerchief. 'Bloody hell, young lady. You've certainly got your work cut out.' In a puff of aftershave, he was gone and his silent, brooding, reticent-to-the-point-of-dumb-stricken audience were staring suspiciously at me.

I dug deep and resurrected my Leonardo DiCaprio technique. Say something unexpected. Provoke a startled reaction. 'Anyone here had thrush?' I asked. 'My intimate bits are itching so badly I feel like using this stick mic as an internal vag-scratcher.'

I had them at 'thrush'. Within seconds, the female redheads were confiding thrush cures – applying live yoghurt to the affected area with a Q-tip was a favourite – recommending knickers with cotton gussets and telling me to lay off the sugar. The chaps were horrified and chipped in sharpish to steer the discussion in a less vulva-based direction. The cameramen sprang into action. The redheads were chatting animatedly. I was listening with a Mother Teresa-meets-Julie Andrews-as-Fräulein Maria expression of rapt attention plastered all over my features. I knew that when the editor got in with the scissors, Multimedia would have no inkling the misty compassion in my eyes was evoked by a description of a nasty rash spreading from front crack all the way to anal fissure.

I was on a roll. The ravishing redheads were rocking. At one point I even reached in, grabbed a small child, put him on my knee and channelled Truly Scrumptious with every fibre of my empathetic, compassionate, listening soul.

The waiting period was agony. Would US bosses anoint me their Oprah? I was back in the cheap end of Hampstead Garden Suburb, almost technically East Finchley, but I kept my head permanently turned to the right and refused ever to look left, so managed to pretend East Finchley was a distant hamlet, not four yards over my left shoulder. I was stuffing a chicken, writing about wet dreams and helping Allegra with a project on sundials. With every passing week, the Oprah dream faded further into unlikely/ludicrous fantasy. No NY Billy Big Bollocks was ever going to speak out of the side of his Groucho Marx cigar and growl: 'Yeah! Vanessa's our Oprah! Sign her up.'

Except that was exactly what happened.

27

Watch Sally Jessy Raphael. Do what she does

I WAS MID-SEPARATING DARKS FROM lights pre-wash. Malcolm rang. 'It's done. They love you. They want you. We want you. We need to meet, agree a fee and sign a contract.'

OMG! You want emotions? Pick one. Relief. Joy. Terror. Amazement. Did I make a fuss and hang bunting from the banisters? Of course not. I kept my excitement under wraps. I couldn't over-jubilate. I didn't want to eclipse the Good Doctor. I was seeking permission to leave my (our) children in his care, plus the au pair's and my mother's, so it couldn't look as if I was truanting off on a fabulous US jolly. I had to imply the trip would be arduous toil. There'd be no element whatsoever of fun – Heaven forbid!

Multimedia put their money where their mouths were. They flew Malcolm, me and my producer Zoe Watson to New York to worship at the feet of talk-show titans Phil Donahue and Sally Jessy Raphael. They gave us the old razzle-dazzle. A stretch limo with a fully stocked bar and television picked us up at JFK. Malcolm flicked on the telly. You'll never guess who was posturing and preening on the screen. *Me!* Of all the zillions of US shows broadcast at that moment, we'd chanced upon a *Maury Povich* programme with me as a guest. I tried to look modest but was chuffed to the core. I was in a stretch limo in the blinking Big Apple en route to Oprah-dom, watching myself on the flipping TV. Come on! Days ago, I'd been smashing out articles on hair conditioner in the dining room for fifty quid a pop. I had to pinch myself. I had to get Malcolm Allsop to pinch me. He didn't look too pleased. *Maury Povich* was made by a rival company. Multimedia were not going to like it one bit.

The learning-from-the-greats idea was better in theory than in practice. I was thirty-two. The US bosses told me to watch a sixty-year-old man and fifty-nine-year-old woman, who'd been hosting their shows for decades, soak up their skill by osmosis and replicate it on film in a studio on the last day of the trip. There was a problem. Although I was a no one and a novice, I knew I was myself, not Phil or Sally, and it would be a monumental mistake to try to copy them. I would never and could never respond to guests or phrase questions the way they did. I was British for a start – less brusque, less confrontational, more humorous and conversational. They were the reigning king and queen, venerated from coast to coast, multimillionaires, and I wasn't fit to lick their boots. I appreciated their superiority. They were legends. I was plankton but I simply couldn't turn myself into a clone of either of them. Deep down, I didn't want to and privately resolved – please pardon the cliché – to be myself.

I developed what we Brits call 'a shocking cold'. Make-up frowned at my reddening nose. Hair – American TV hairdressers rule the roost in a way our modest crimpers wouldn't dare – snipped several inches off the front strands of my hair without asking permission. If they had I wouldn't have allowed it. Talk about Alexander Pope's *The Rape of the Lock*. It was too late – no use crying over cut hair. It took every ounce of resistance to stop them chopping off the lot and forcing a horrible American hair helmet on me. Queen of Wardrobe Cynthia Raffa-Rickman produced supersized silk trouser suits in 'celadon, aqua, double cream and espresso'. Americans powdered, curled, hemmed, twirled and commented on me.

I was strictly instructed never, under pain of death, to touch my face or hair while broadcasting. Note to Richard Madeley of this parish, famed for fiddling with his nose on camera. I was commanded never, ever, to touch my clothes. 'You are dressed. That's it. Do not revisit.' I was taken to dinners I was too bunged up with the cold to eat, photographed with Phil Donahue, carted in and out of enormous boardrooms.

First mention of Jerry Springer

Meetings were often interrupted by panic-stricken individuals running in and sighing: 'Jerry!' Feverish calls ensued. 'Jerry!'

'What the fuck has happened now?'

'Tell it to me straight. Jerry! Fuck!'

Information filtered through. The Jerry, of exclamation marks, was a bloke called Jerry Springer. He'd been Mayor of Cincinnati, paid for a prostitute's services by cheque and was sent packing. Multimedia was trying to mould this perspicacious politician into a talk-show host. In 1994 it wasn't going well. Jerry was making serious political discussion shows no one wanted to watch. We all know what happened next. Jerry slid spectacularly downmarket, and the show became one of the most successful in the world. Hair sprayed solid with Elnett, I made it through the final filmed recording. I didn't try to be Phil or Sally. I listened intently, spoke politely.

On the eight-hour flight home Malcolm, Zoe and I were too talk showed-out to speak a word.

28

Why make one pilot when you can make twelve?

DON'T ASSUME THE DEAL WAS done. I'd passed the can-she-listen? test. The Americans were on board, but there was a major obstacle blocking my Yellow Brick Road. Anglia hadn't yet been commissioned to make the show. The rival Oprah-hatchers were closing in. We needed Marcus Plantin, ITV head, to love our programme, not theirs. We had to win his heart.

Luckily, Malcolm Allsop was a man with a cunning plan. We would shoot twelve episodes of *Vanessa*. They would be shown locally in the Anglia region. Making not just one pilot show, but effectively twelve pilots would give me a chance to get better at presenting and him more shows to choose from. We could present Plantin with the best of the bunch. I'd never seen anyone pitch a new TV show before, so at the time I didn't appreciate what an innovative and resourceful boss I'd landed in Malcolm.

Why I didn't want an agent

I was invited to dinner at the ultra-stuffy posh RAC club at St James's to be pinned down financially by Malcolm and a gentlemanly grandee from the Anglia board of governors. Most budding TV hosts would have found a showbiz agent with a Rottweiler reputation. I decided not to. I've always believed I could do a decent job on my own behalf and save paying an agent 20 per cent of everything I earn. I've rarely if ever used an agent since, and never for any of my long-running jobs. I can't stick the idea of some fellow rolling over in bed, spooning his wife and while he snoozes in conjugal comfort, earning 20 per cent of whatever I'm being paid for schlepping out to work. Agents' fees go on for ever.

The 'talent', which is what presenters and actors are called, very occasionally get a job because their agent does a hard sell and lands them a contract. Gigs usually come in because someone rings your agent and says: 'Can we have Vanessa Feltz for *Strictly*, please?' The agent did nothing to generate the call. Yet, because you're signed to him, he pockets 20 per cent of your fee. If that call results in a year's contract, the agent is paid 20 per cent of your earnings every week for a year. If the contract lasts ten or twenty years, the agent still trousers 20 per cent of the total. What? Twenty years of bumper cash, just for taking a phone call? Come off it. The whole thing stinks. I wasn't up for that arrangement then and I'm not up for it now. I'm not scared to talk about money. I don't find it embarrassing. There is, however, always the niggling fear that you might chronically under- or overestimate the ballpark figure and shoot yourself in the foot.

Anyway, that night I was brimming with fake confidence and when they named their figure over the smoked haddock, I calmly asked for time to consider it and left the dining room. I stayed in the vestibule, cooling my heels for much longer than was comfortable. I wanted the suits to be discombobulated. I hoped they'd fear I considered their offer so pathetic I'd stormed out in disgust and was on the 26 bus heading home. I wasn't. I was just biting my nails, pacing the floor, watching the clock and waiting for the right moment to stride back in, head held high and demand double. Would you believe me if I told you I got it?

29

Brits can't make talk shows

WE HAD TWELVE *VANESSA* SHOWS to make. There was no precedent. This was virgin terrain. I didn't know how to be a talk-show host. The audience didn't know how to be a talk-show audience. The guests didn't have the faintest idea how to be talk-show guests. Anyone who was anyone in television had strident views about the project: 'It's culturally impossible'; 'A British talk show? Don't make me laugh'; 'We British are a stiff-upper-lip, proud, private people. We're not, Heaven forfend, American'; 'Washing our dirty linen in public? We would never descend so low. You're thinking of making a talk show? Here? In the United Kingdom? Good luck with that. Don't hold your breath.'

For a while I thought the nay-sayers might have a point. Our guests frequently bottled it. They'd be booked to say: 'I'd never shag a woman bigger than a size twelve.' They'd be bouncy, brash and bolshie on the phone. The researchers would promise me the sun, moon and stars. 'You'll love him, Vanessa. He's a grade-A male chauvinist pig. He's a show-off, ready to face down angry ladies in the audience.' The fellow would pitch up snarling. He'd be champing at the bit in the green room. I'd whirl in for a pre-show handshake. He'd be oozing macho swagger. The signature tune would blare. I'd announce his name. He'd sally on to the set, take one look at the audience and – right there on the spot – change his tune completely. 'Derek here wouldn't dream of rumpy-pumpy with any woman larger than a size twelve,' I'd declare. 'Derek, why have you ruled out all the voluptuous ladies in our audience today?'

'I haven't, Vanessa,' he'd reply sweetly. 'These ladies are absolutely stunning in all shapes and sizes, and I'd be a very lucky man to get a snog off any of them.'

Cue consternation in the gallery. Cue an apoplectic Malcolm Allsop, tearing a strip off the researchers. Cue gentle persuasion from me as I tried to coax the guest back to his previous position. Cue confusion among the audience who were being persuaded to boo him, but now saw he was just a pussy cat beneath the tattoos. Cue the chastening realization that this talk-show lark is only easy when everyone involved gets it and has the guts and gusto to play their part. The guests must not switch personalities or stories mid-recording. The audience must not sit in silence being politely non-judgemental, and the host must be versatile enough to shift focus and present an entirely different programme from the one she expected without so much as batting an eyelid.

Show Three

Show Three stood out a mile. You'd think I'd remember what it was about, wouldn't you? Unfortunately even Google has forgotten. I should have kept a diary. Everyone agreed it might have the magical ingredient to make it the Plantin Persuader.

Malcolm struck again. He wasn't prepared to stick the tape in the post with a covering letter. Instead, he had a pal who had a pal who knew someone who happened to live near the Plantin homestead. He was prepared to make a pilgrimage to Plantin-ville and pop it through Marcus's letterbox personally. He did so. We waited. Meanwhile the programmes were being broadcast in the Anglia region. We could have driven in the direction of Thetford, knocked on a random door and asked Mrs Jones if we could please settle on her sofa for forty-five minutes to watch the programme going out in real time. We didn't.

Had Marcus Plantin opened the brown envelope? Had we captured his heart, hunger for a hit and open ITV wallet? Had he chucked Show Three into the wheelie bin in disgust? Had Mrs Plantin written a recipe on the envelope and tucked it away in her dinner-party cordon bleu drawer never to be seen again? Silence. Silence. Deafening silence.

30

All systems go!

On 15 July 1994, we moved house again. My mother and mother-in-law were unpacking boxes of my kitchen items and driving me insane. Each time they pulled out a battered colander or lidless saucepan, they lectured me on the shamefully shabby state of my cooking utensils and expressed amazement I'd managed to keep the children alive without a full set of matching Le Creuset. The house had good bones. It would only take a bumper injection of money and a dollop of imagination to restore its 1930s becolumned glory. Alas, we were going to have to live with the lime-green carpet, citrus bedrooms, antiquated kitchen and sixty-five-year-old plumbing for ever. Paying the mortgage would take every penny we earned. Sitting stock still on orange boxes with a bowl of gruel each would be as much fun as we could afford for aeons to come.

I didn't mind. It had taken us nine years of grafting, moving and doing up three houses on a shoestring to fetch up in this Finchley N3 haven of loveliness. It was called 'The Gables'. I wanted to change it to 'The Baygels'. The idea was instantly vetoed by husband, parents and mother-in-law.

I whistled while I worked, dreaming of a happy, healthy future in this terrific house when to my surprise – it was a new number (landline, of course) and I hadn't given it to anyone yet – the phone rang. It was Malcolm. 'Good news. Marcus Plantin loves it. ITV has commissioned series one of *Vanessa*. It will run on two weekday afternoons for twelve weeks. All systems go. You need to go to New York next week to hobnob and pick up twenty-four suits. Can you fly on Tuesday?'

Could I? Even better, could I scoop up the Good Doctor and the girls and give us all a quick holiday we could never have afforded in a million years? I couldn't believe it when he said, 'Yes!' In one phone call we'd gone from impoverished prisoners of a crippling mortgage to a family packing for a flight to New York and four whole nights at the absurdly posh Helmsley Palace (owned by Harry, husband of Leona Helmsley, who ended up in prison for federal income-tax evasion after her housekeeper testified to hearing her say, 'We don't pay taxes. Only the little people pay taxes.').

How not to handle being interviewed

Builders bashed out the old kitchen and pulled up the acid-green carpet. We could afford to pay them. Bliss! ITV set up a raft of interviews. Did I want them to book a London hotel? No, thank you. The girls were off school for the summer. It would be simpler for the journalists to see me in my natural habitat, at home. Fleet Street's finest picked their way through the rubble. Seasoned hacks from *The Times* and the *Guardian*, showbiz reporters from *Woman's Own* and the *Sun* trekked to Finchley.

Naturally I offered beverages and biscuits. 'What would you like to drink?' I asked a well-known female journo. 'Tea, coffee, water, Diet Coke, orange juice, Aqua Libra?'

'Oooh, I love Aqua Libra, my favourite,' she replied. Her article, in a broadsheet newspaper, began, 'Vanessa Feltz poured me a glass of urine-coloured liquid.'

Inviting journalists to your home is a rookie mistake. No PR guru at ITV advised me against it. They should have done. I didn't know it then, but the key is not to arm your interrogators with any extra ammunition. See them in a neutral location. Wear something beige and nondescript. If you normally festoon yourself in jewellery or spray yourself in fruity fragrance, don't. Resist swigging cocktails. Order a glass of water or a cup of coffee. Offer to pay for their lavish lunch even though you know they claim it back on expenses and you don't. Never, for an instant, believe they are your chum. Don't confide information you should keep private because you hope it will make them like you. Don't utter the phrase 'This is off the record', even if you've heard people say it in films. If you say something, you've said it. You can never unsay it. It's open season. 'Off-the-record' comments will be back to haunt and taunt you – and don't believe that 'yesterday's fish-and-chips paper' guff. Whatever tosh is written stays on Google to torment you till all eternity.

July and August 1994 were downright weird. I wasn't famous. No one, except Anglia dwellers, had seen the programme, yet containers of TV clothes were being shipped over from the States and I'd signed a contract for a network show.

31

Critics are disgusted

Vanessa launched in September. Critics were disgusted. Nothing in the history of British broadcasting had prepared them for this avalanche of vulgarity. They were offended, appalled and insulted. Their sensitive constitutions were affronted. There could be, they decreed, no appetite in this green, pleasant and polite country for such fare.

They were wrong. Within a fortnight viewing figures for *Vanessa* surpassed the numbers for the imported *Oprah* show. The self-proclaimed intelligentsia would rather have their eyes poked out than sully themselves by watching the 'flow of sewage' we were producing. The daytime audience loved it.

The hoo-ha was absurd because the programme was, in fact, tame and gentle. It often featured eccentrics like 'Britain's Meanest Husband' and 'Cone-Man the Barbarian', who collected traffic cones, or hopeless romantics: 'I'm In Love With My Next-door Neighbour'. We were governed by suffocatingly stringent broadcasting restrictions. Lie-detector test results and shock guests announcing titillating revelations were notching up big business for Jerry Springer. Multimedia spectacularly turned around his failing show by dragging it juicily downmarket.

In the UK having people stalk on set and announce, 'I'm not

your father, I'm a woman!'; 'You think I'm your son but I'm your brother's child!'; 'I'm pregnant by my father-in-law and I'm not saying sorry!' was strictly forbidden. We couldn't even allow the rampant exhibitionists begging to give their naughty bits an airing to be semi-naked. We were allowed transvestite husbands and their loving wives chatting about how they swapped fashion tips and plucked each other's eyebrows, but nothing saucier or more salacious. *Vanessa* was, if you bothered to watch it, genteel enough to enjoy over Victoria sponge with the vicar. We were seaside postcard not porn. Of course, the cascade of criticism didn't do our 'nudge-nudge, wink-wink' reputation an ounce of harm. The more the cognoscenti despised the show, the more the audience approved.

After just one series, ITV commissioned an extra show each week. We were now turning out three *Vanessa*s. We were in our stride. Producers and researchers learned to book and brief. I learned to fire up the throng and rev up the guests. The audience learned to ditch their inhibitions, and the guests stopped losing their nerve and started firing on all cylinders. We made sad and touching shows. We made suggestive and hilarious shows.

Posh people still didn't get it. 'You were sexually abused by your mother. Why on earth would you want to go on television and talk about something so private? What is it with these people? It's utterly mind-boggling.' It wasn't bewildering to me or the *Vanessa* team. People wanted to bring their traumas and tribulations to my programme because they had been through hell and badly needed to be heard. Psychotherapy is expensive privately and incredibly hard to access on the NHS. People have an intimate, personal relationship with the presenters they allow into their living rooms. You can see it clearly on Channel 4's mega-sensation *Gogglebox*. Viewers feel a genuine connection with presenters they like. The chance to talk to them in real life on a television programme they enjoy feels like a wonderful opportunity.

Telling their story in front of me and my studio audience was cathartic. They were listened to – sometimes challenged – and supported. Often, if they'd been living under a crushing burden of guilt or shame, they experienced a comforting kind of absolution. The head TV honchos wouldn't have watched the show if you'd paid them. (Hang on. They *were* being paid to do exactly that but didn't for fear of contamination.) They would never have entertained the idea of parading their peccadilloes on TV. They couldn't grasp that, for some people, appearing on *Vanessa* was exciting and even, in some cases, healing.

32

Overexposure, so soon!

WE'VE ALL SEEN IT. A TV 'personality' appears out of nowhere and suddenly they're plastered across everything. It's beyond a joke. Their face is stamped on your butter and grinning up at you from your duvet cover. Their arse is beamed on the Houses of Parliament. 'Overexposed!' we yell at the telly, as the dude bobs up fronting a game show, trilling out of tune on *The Masked Singer*, doing a cameo in *Bridgerton* and flogging online bingo. We wonder why they don't do us all a favour once in a while and stay in quietly with a biryani.

I'll tell you why. They're in a daze. They're grappling with the first flutterings of fame. They've no idea if it will last. They've never been offered shedloads of cash before and they can't guarantee it will ever happen again. They invariably have a pushy agent hoovering up the gigs and shoving them towards the money. Every offer is an offer they can't refuse. You must be rich and established to turn down highly paid work. Overexposure is a risk that seems 100 per cent worth taking. Naturally, I took it.

My mother was diagnosed with endometrial cancer when she was fifty-three. These were my thoughts in order:

1. Please, please, God, make her better. Don't let her die.
2. My poor mother. She must be terrified.
3. I won't eat. Ever again. God, I'll do a deal with you. Make my mother better and I'll stop eating. I promise. I swear. I solemnly vow. She's desperate for me to be thinner. I love her with all my heart. I want to give her as much joy and pleasure as I can to show how much I love her. If I get thin, she'd be happy.

What kind of daughter hears their mother has cancer and doesn't do everything they can to cheer her up? It's simple. It's the thing she wants me to do most. I'll do it. Frankly, why would I want to eat anyway, with a sick mother and a big gristly lump of fear blocking my throat?

You might have thought anxiety would annihilate my appetite. Who tucks in to banoffee pie with gusto when their mum's on a drip in intensive care? Heart-breaking news is the ultimate weight-loss incentive. We'd been told my mother had ten months to live. She didn't know and didn't want to know. This was the equation: Vanessa plus food equals even more misery for your dying mother. Vanessa plus empty plate equals possible smiles from your mother just prior to her premature death. Easy-peasy. Step away from the supper. Banish breakfast. Jettison jam roly-poly. Channel *Oliver Twist*. I wanted to starve myself to a silhouette. I wanted to be a hologram. I failed. I managed to shake off seven or eight pounds, but the scales stuck. I was still eating. I detested myself.

Ten months later Valerie Feltz was alive, playing golf, carting my father off to tour the stately homes of England and making such a convincing fist of being healthy even the doctors began to forget she was on borrowed time. She was still hosting Friday-night dinner. She and my father were still taking Allegra and Saskia to see amateur-dramatic productions of *Carousel* and *Guys and*

Dolls at the Millfield Theatre in Edmonton. She never acknowledged it could be any other way. She was still alive and still fiercely critical when *Vanessa* launched in 1994.

The official PR pictures, snapped on the roof at ITV in bright sunshine featured my smiling surgeon husband, the two girls aged eight and five, with Pre-Raphaelite waist-length blonde hair wearing matching floral French Provençal pale blue dresses and me in a scarlet size eighteen American TV suit, with huge nineties earrings, high heels and a five-star blow-dry. I looked exactly what I was: thirty-two, excited and optimistic, on the cusp of career adventure, surrounded by my family. I thought the pictures captured unalloyed happiness.

My parents were on holiday when they featured in the *Daily Mail*. Copies reached Majorca/Marbella, or whichever resort they were frying their faces in, a day later. They rang to tell me they hated them. I looked huge. They loathed the girls' dresses. My grin was inane. They were worried about all the publicity. It could implode and drag the entire family into disrepute. Did I know what I was doing? Did I realize how I was looking? If ever there was a moment to lose weight, this, surely, Vanessa, was it. And, while they were on the subject, wasn't the whole having-a-TV-programme-with-your-name-on-it thing vulgar, attention-seeking, not quite *comme il faut*?

My dad imported the first edible panties to the United Kingdom. They were pineapple flavour. Was that quite *comme il faut*? I never asked that question. I didn't say a word. Crestfallen doesn't begin to describe my feelings. Neither does, bubble burst, confidence nuked. I was so voraciously hungry, I had to make an urgent trip to Sherrards bakery for an emergency doughnut.

33

They want your autograph? Don't be ridiculous

Afternoon shows are hardly primetime. I wasn't Mrs Saturday Night. I could easily have escaped notice. I might have been lost among the folk who bob up on the small screen never to be heard of again. I wasn't. I still don't know why. Instead, I was instantly famous. My parents weren't keen. 'Vanessa, why is that man staring at you? Move swiftly on. Don't give him the time of day. He's a *schmendrik* [imbecile]. What does that woman want? Your autograph? Tell her not to be absurd. How does the Good Doctor feel about this nonsense? He's a surgeon. You are just showing off on daytime television. Whatever you do, don't upset him. Allegra's headmistress wants you to go and give a talk at Haberdashers'? What for, for Heaven's sake? That school has churned out lawyers and scientists. What on earth do they want with you? You've been nominated for best daytime show at the National Television Awards? Trevor McDonald is going to announce your name? Sir Trevor McDonald knows who you are? What a waste of his time. You've been asked to St James's Palace to attend an event hosted by Prince Charles? Why? What does he want *you* there for? What does the Good Doctor think? It's the au pair's night off. Who is going to look after the children?'

My parents revered the late Chief Rabbi Lord Jonathan Sacks.

He had a triple first from Cambridge, was a philosopher and spoke in an English accent, not the middle-European intonation of previous chief rabbis. My parents thought he was the rabbinical bee's knees. Lord Sacks approached them at a function. I'd interviewed him frequently for GLR's *Jewish London*. 'Mr and Mrs Feltz?' said the Chief. 'You must be so proud of your Vanessa.' Norman and Valerie were never again quite such ardent Lord Sacks fans.

Tabloid fun and games with my weight played perfectly into my parents' hands. Seeing my ample *toochas* (rear end) plastered over magazine covers confirmed not only that my mother had been right to diet me but that I had been wilfully foolish to flout her restrictions. Female journalists were outraged by my size. I hadn't developed a rhinoceros hide. It was hurtful as heck. I had nowhere to turn, except the fridge. Right there in the public eye, in plain sight, on the television screen, I was getting fatter every day.

34

Welcome to the dead mothers society

My mother started dying. I know we'd been warned but that was three years earlier and we'd stopped believing it. It took six weeks. One minute she was getting a par four on the eleventh hole at Potters Bar and dancing in pale pink Pauline of Cockfosters taffeta to 'I Just Called To Say I Love You' at Sharon Anisfeld's second wedding, the next she was near death. She was fifty-seven. I was thirty-three. I loved her immensely and intensely.

I was making *Vanessa* in Norwich and living in Finchley. The girls were nine and six. I was a columnist at the *Mirror*, presenting *Value for Money* on BBC1 and a show on Talk Radio every Sunday morning. The only place I wanted to be was at my mother's bedside. I insisted we record three *Vanessa*s a day. ITV said it couldn't be done. I knew they did it in America. They tried it. We did it, and I dashed back to London after each taping. Sometimes my mother was in hospital, sometimes at home, palpably sickening, dwindling, skin sunken, yellow, fading away.

I was slumped with sorrow by her hospital bed. My mother's eyes were closed. She was too weak to sustain herself with her adored Keats and Shelley. I loved her so much I wanted my blood to flow from my body to hers and keep her alive. There ought to be an adult umbilical cord through which we can funnel love and energy into our dear ones when they need it. I loved her so much I couldn't bear to breathe because she could barely breathe.

The television was on. We took no notice. My face popped up. It was episode one of series two. That afternoon, the concept was light-hearted. There'd be a ring at my TV door. I'd answer it and a lookalike would enter the studio. I'd pretend to be fooled. Mischief would ensue. There was a Norman Wisdom, a Joan Collins and Elvis – you can imagine. It was silly flotsam to tickle the audience. It wasn't meant to be Strindberg.

At first my dying mother watched in silence. Then she mustered every ounce of energy and let me have it. 'What are you supposed to be wearing? That suit doesn't button all the way down because you're so fat. It gives up the ghost where your waist ought to be. What were they thinking of letting you wear it? The wardrobe department should be fired. And who does your hair? Can't you tell them they need to smooth out your frizz? The lighting is dreadful. And those shoes? A disaster. Stumpy. Unflattering. Shortening your legs.'

Nothing was spared. My dying mother assassinated the guests, the audience, the signature tune, even the acrylic carpet 'such a cheap, nasty look'. Professionally barbed spiteful critics with reputations for malevolence exude a small smattering of mercy. My mother was merciless, poisonous. I was so stung I couldn't even cry. I had to leave to collect Allegra from school. Driving to Elstree, I found I was patting myself in a soothing there-there motion. It must have been a subconscious way to cope with feeling so hurt.

On 20 October 1995 my mother, Valerie Joy Feltz, died. She died at four a.m. and earth was shovelled on her coffin at two p.m. the same day. Gentiles punctuate life with alcohol. 'What's your poison? Get in the bevvies.' Jewish people eat. Food is Jewish life. It's also Jewish death. Would you believe there's designated death food? Returning from a funeral, mourners are required to eat a plain bagel – the circular shape symbolizes continuity – a hardboiled egg, same reason, and a piece of herring, because the salt reflects the tears shed. My mother died and was buried so fast I wanted to exhume her to check she wasn't still breathing. I wanted to fling myself alive on her funeral pyre and be barbecued. Leaving her all alone in Bushey Cemetery when all her friends and family were sobbing and eating poached salmon in my house in Finchley seemed unkind. I thought she'd be lonely with everyone she knew miles away. I wanted to go and sleep on the freshly shovelled soil. I wanted to get into a red convertible and drive to the South of France without a word to anyone.

My mother was dead. She was fifty-seven. I had no idea what to do. I was certain of one thing, though. I'd choke on that bagel-egg-herring combo. I was half an orphan. I'd seen the burial-society blokes carrying out my mother's corpse in a body bag only eight hours earlier. How could I have any appetite? Food? Forget it. I couldn't possibly eat a thing. Nothing at all for me, thank you. I couldn't stomach it. Definitely not. But what is that? Amanda Huberman's fried chicken? With potato salad? Well, maybe just a tiny bit. And Carole Landau's special rice with apricots and almonds? Half a spoonful. I couldn't believe my behaviour.

My mother was dead. Do you hear me? I had a dead mother. Obviously, I'd be too wretched to eat. Why would I want to? Grief subsumed my hunger. I was so bereaved I was never going to eat again, not ever. What's more, I knew my mother was watching me in Heaven – or wherever she was – and she'd see me not eating. She'd be pleased with me. She'd be proud. Only I'd just polished off Amanda's chicken, Helen Bloom's pasta ring, Carole's special rice and a dollop of Deborah Hofman's trifle. I was ravenous.

35

She's fucking huge. What can we dress her in so she doesn't frighten the horses?

PROVING THE COGNOSCENTI DIDN'T HAVE a clue, *Vanessa* was a soaraway success. Scheduled first for two weekday afternoons, then three, ITV couldn't ignore sensational ratings. Three years later, amid no fanfare whatsoever, they moved us to five mornings a week, airing before Richard and Judy on *This Morning*. *Vanessa* scooped a 43 per cent audience share: of all the people watching TV at that time of day, 43 per cent were glued to 'I'm a Bride Who Had Sex With the Best Man!' I was dashing up and down to Norwich to film six shows in two days at Anglia TV, trying to console my girls for the loss of their beloved Grandma Valerie, writing a column for the *Mirror* and self-soothing with just a few too many Revels. The weight was piling on. My Mystic Meg dead mother was right. I was trying to fill an emotional chasm with Cadbury's Dairy Milk.

Time to tell you about Fat People Fashion in nineties Britain. There wasn't any. OK, there was Evans, formerly known as 'Evans Outsize', bristling with polyester tents and drawstring-waist trousers. Dawn French, the darling, valiantly attempted to address the drought with her company Sixteen47, but her aesthetic was art teacher meets fossil hunter. The clothes were beautifully made but not glamorous. If you were over a size eighteen, and I was a twenty edging towards a twenty-two, you might as well take a leaf from Julie Andrews's book in *The Sound of Music* and make an outfit from the living-room curtains.

Designers adore wraith-like models. The closer your figure is to a clothes-hanger the better garments drape on your body. No fashionista wants to dress a fatty. Buyers believed fat women were so busy eating all the pies they'd blow their budgets on food, not clothes. Manufacturers had woken up to the purchasing power of the pink pound, but no one thought it was worth wooing the fat fiver. If you were over a size sixteen fashion was dead to you. You had forfeited the right to style or elegance.

In 1995, BBC1 flew me to New York for their consumer programme *Value for Money* to investigate 'fashion for the fuller figure' stateside. I was Augustus Gloop in Willy Wonka's chocolate factory. Voluptuous American women could buy jeans, strapless bras, pencil skirts, off-the-shoulder dresses, hot pants, negligées. US stores didn't stigmatize women with abundant figures. Their designated plus-size departments were welcomingly entitled: 'Woman'. I needed a hot and cold running supply of brightly coloured nineties power suits for *Vanessa*. The UK was a bare cupboard. The US mavens decreed I must have a new suit for every episode, no repeats allowed.

Cynthia Raffa-Rickman was enlisted to scoop up four hundred fat-lady TV suits a year. She hit the ground running. Sprinting into her atelier with machinists, assistants and assistants' assistants she kissed me warmly and yelled: 'We have the three-piece in chartreuse, crème de menthe, vanilla, banana and eau de Nil. I've pulled the velvet in aubergine, acid green, margarine, Benzedrine. Slip on the de la Renta, the Herrera, the Kmart and the Donna Karan. Oh! My! God! You look adorable! You're a peach!

The carmine is incredible with your eyes! You're a doll! Santos, Mercedes, Alvin, Granroy, Jesus, wrap and pack, wrap and pack. This lady is not going to be naked.'

At home, at least and at last, I was off the hook. My father copped off with some woman in the six weeks it took my mother to die. When she balked at going away for the weekend with him six days after his wife's (my mother's) funeral, he calmly replaced her with some other superannuated broad and flitted off into the wide blue yonder. The Yiddish chorus was almost silenced. I could slurp a Cornetto without commentary. I could almost begin to relax and savour the creamy bliss dissolving on my tongue. Who knows? If I'd been left to eat in peace, I might even have reached the state of nirvana where I could have put the ice cream down half finished and written a poem instead. I never had the chance to find out.

36

The Big Breakfast

TRACEY MACLEOD AND NICHOLAS LAZARUS from production company Planet 24 came to see me. How would I feel about boarding the *Big Breakfast* bed? The bed upon which Paula Yates came clinically close to having actual sexual intercourse with Michael Hutchence as the nation munched its collective Frosties in envious awe? The bed upon which Lily Savage swapped sensational secrets with Catherine Zeta-Jones? The bed upon which every British breakfaster spooning in Ready Brek yearned desperately to cavort?

What do you think I said? 'Sorry. No can do. I'm washing my hair'? My mother had been dead only five months. My father had disappeared over the horizon with a dowdy divorcée. I was thirty-three and drowning in grief. *The Big Breakfast* was my lifeboat.

I was invited to a swanky breakfast (appropriate) at the Savoy Grill to seal the deal. 'We're so excited to have you,' said a spokesman. 'We're really excited, so excited you can have anything you want. I mean anything: drugs, sex, anything.' I thought about it. 'How about somebody to do the ironing and some Haliborange?' I replied.

What a prig and plonker I was. I should have demanded sex with a well-hung Chippendale and a mountainous cocaine chaser after every show. I was never offered a chance like that again, damn it! Unlike Edith Piaf, *je regrette* absolutely everything.

The Big Breakfast doubled Cynthia Raffa-Rickman's workload. Paula was petite. Lily was a fella in a frock. The bosses didn't say to my face: 'Vanessa. We're concerned. Your acres of flesh flowing over the bed might put the nation off its Weetabix. How the bloody hell are we going to dress you, so you don't look like a dog's dinner even the schnauzer won't touch?' They were too nice. They still are. I'm still friendly with all three Planet 24 maestros to this day. One of them is actual Bob actual Geldof, for goodness' sake. Sir Bob of Band Aid doesn't go round telling fat women they make him want to puke. They didn't exactly say it, but their concerned expressions were a clue. How could they doll me up, so I didn't frighten the horses?

The answer, dear Reader, turned out to be evening gowns. If they swathed me in full-length chiffon, tiers of organza, crinolines in emerald satin at breakfast time when the country was climbing out of its pyjamas, I'd look so startlingly eccentric, my size would be a bagatelle. I'd be befeathered, bejewelled, bedecked in sequins, layered in lace. Sartorial convention would be turned upside down. I'd be satirizing myself. Were there enough size twenty-two ballgowns in the British Isles to satisfy our demands? You know the answer. Cynthia Raffa-Rickman was charged with roaming the Land of the Free, hoovering up all the XXL Halston, Gucci and Fiorucci she could find.

37

Go to work on a bed

IMAGINE GOING TO WORK EVERY morning and your job is climbing on to a pretend bed with Charlton Heston, Jim Carrey, Andie MacDowell, Alicia Silverstone, Johnny Mathis, Graham Norton or Goldie Hawn. Imagine getting glammed up in full make-up, false eyelashes, cascading ringlets and a silk gold lamé fishtail ballgown at seven a.m. Imagine wafting down the stairs to Lock Keeper's Cottages, kissing Zoe Ball hello, waving at Joe Mangel of *Neighbours*, aka Mark Little, having a laugh with the Family of the Week, patting a couple of pecking chickens on the head, gossiping with the Bay City Rollers and joining the late great Joan Rivers on the bed? It wasn't work. It was joy, sheer broadcasting bliss. I couldn't believe my luck.

Chaos was king. The programme was built on it. There was no autocue. Presenters carried cards, usually upside down or in the wrong order, and had to make up the script on the hoof. No one could remember the phone number. That's why they invented the iconic line: DON'T PHONE, IT'S JUST FOR FUN. (Imagine saying that now when TV companies are desperate to wring every penny out of the audience by hooking them into ringing lucrative phone lines.) We ran everywhere: across the canal, into the garden, up and down the higgledy-piggledy stairs, vaulting over the tangled jumble of cables. Cameramen tripped and fell in the lock. Guests tripped and fell down the stairs.

Presenter Richard Orford was abseiling down a city scraper. Footage stopped abruptly when the screen went grey as something crashed downwards and hit the pavement. Was it Richard? Was he the splat? Was he dead? They cut to me. I filibustered, filling time with fluff and nonsense till someone handed me a scribbled note. 'Cameraman dropped camera. Richard fine. Camera toast.'

The Lazari, Nicholas Lazarus and Sheldon Lazarus, astoundingly not related, were the international celebrity booking team from Heaven. The two enticed the most dazzling A-listers in Hollywood's constellation all the way to the backstreets of Stratford, East London. The world's twinkliest stars arrived at the *BB* jet-lagged, sometimes wired and always dazed and confused. None of them had ever seen anything like it. This tiny artisan's dwelling wasn't a studio. A minuscule kitchen with giant 3D fried eggs on the walls, someone's collection of erupting papier-mâché volcanoes and four live rabbits nibbling presenters' ears didn't feature in their broadcasting repertoire. You could see some of them making a mental note to give their agent a post-show bollocking.

By the time guests pitched up on the bed with me, they didn't know if they were coming or going. My two blessed predecessors are sharing a cloud in Heaven, so you'll have to rely on my testimony when I tell you broadcasting on a bed does funny things to people. Some fall asleep. It's a reflex action. They're lying down. There's a pile of comfy pillows. Hey presto, zzzzzzz! Some become amorous. They're lying down. There's a bosomy blonde within arm's reach. They feel a boner coming on and, regardless of cameramen, director and live transmission, they're in the mood for urgent rumpy-pumpy. Whichever genius dreamed up the bed – I suspect it was *Survivor* creator and co-Planet 24 boss Charlie Parsons – had a cunning plan. The bed would be a disinhibitor.

The unexpected informal environment would jolt celebrities out of their comfortable rut, stop them telling their usual made-for-TV stories and prompt something genuine for a change. It worked a treat.

Here are a couple of things about the *BB* bed you may not know.

1. It was hollow. It had to be so when Miss Piggy, Kermit and Gonzo appeared above the duvet, their legendary puppeteers could climb in below and conceal themselves.
2. It was transportable. If a cosmic superstar couldn't find the time or inclination to go to the bed, the bed would go to them.

A portable bed was kept in a van for emergencies and would be driven to the London hotel (usually the Savoy or the Dorchester) at which the star's press junket was in full swing. Interviewers from all over the world would form an orderly queue, enter the room for their allotted five minutes, have a mic clipped on and do their best to winkle a memorable conversation from said superstar in a production line of identical chats. *The Big Breakfast*, however, held sufficient sway to extract superstars from the junket and install them on our bed, which had been hastily assembled in one of the bedrooms. Superstars didn't know why they'd been manhandled down a corridor to lie on a bed with a size twenty-two blonde, which made their answers to my impertinent questions even more authentic.

Presenters should hang their heads in shame for cranking out books about their interviewees, as if ten minutes in their company has given them valuable insight into their subject's psyche. You don't get to know the nitty-gritty about a Hollywood A-lister in the space of a single interview. You do, however, get the right to form your own opinion of their politeness, personal hygiene and fanciability.

38

Damn! I should have shagged Wesley Snipes

Woody Harrelson

ECCENTRIC, POSSIBLY STONED, IMPOSSIBLE TO sustain a logical conversation with, Woody Harrelson distinguished himself by being the only on-the-bed guest to peel a mango meticulously throughout the interview without referring to it. I was buggered if I'd give him the satisfaction of mentioning the damned fruit, so I didn't refer to it either. Harrelson peeled, and mumbled, mumbled and peeled. I found him difficult to love but not as difficult as

Dennis Quaid

who removed his shoes and socks mid-chat without explaining why and reclined, wiggling his hairy toes in a belligerent simmering funk. He didn't bother to hide his longing to be anywhere else. At the time, I felt sorry for poor Meg Ryan. What had the winsome star of *When Harry Met Sally* done to deserve being hitched to such a misery guts? Meg Ryan gave Michael Parkinson the silent treatment in an agonizing interview in 2003. She and Quaid (since divorced) deserved each other.

Joan Rivers

Joan was ridiculously quick-witted. Whatever I asked, she shot back an instant hilarious one-liner. She was so slick it looked as if we'd rehearsed. This is how interviews with comics usually work. Their people supply your people with a list of cues. You memorize it and say, fake-casually, to the famous comic: 'I hear you keep a gerbil?'

They reply, fake-spontaneously: 'A gerbil, Vanessa. Funny you should ask me that.' Then they tell their gerbil joke. That's how it's done. When Joan was aboard the bed, people thought we'd been working on material for weeks. We hadn't. *The Big Breakfast* wasn't about seamless production values. Joan and I had never spoken till she pitched up as the principal guest at our *Big Breakfast* wedding. We both wore huge hats with floral trims. She was barbed and brilliant about brides, grooms, mothers-in-law, secretly gay fathers-in-law, best men copping off with bridesmaids. Her gags came so thick and fast you couldn't see her brain flicking through her mental Filofax to unearth each joke.

She was in the UK shifting huge quantities of costume jewellery on shopping channel QVC. Most stars of her stature wouldn't have condescended to grace such an outlet with their presence. Joan was merrily making millions. She took a liking to me and gave me a fat fake-gold diamanté-studded brooch shaped like a bee. Its body clicked open to reveal a watch. I loved her but I didn't like the brooch. I think I donated it to a raffle or something. I wish I hadn't. Joan is gone. How I'd love to flick open my bee brooch now and check the time in her memory.

Goldie Hawn

We all know Goldie Hawn is beautiful, but did you realize she is also warm, wise and wonderful? There's no need for A-listers to give humble British interviewers the time of day. Plenty march in, say what they have to say and march straight out without even a goodbye. Why should they bother being friendly? They've publicized the film, play, tour, album or series. They don't need to make nice. They can't be arsed to wish a nonentity a good day. They're ticking off the obligatory gigs in their contract. Why would they smile, nod or say hello?

Stars are people. Some people are naturally polite and friendly. Goldie Hawn is one of those people. She's an international megabrand. She hardly needs endorsement from a Channel 4 morning audience, but the moment our conversation finished, she asked: 'Was I OK? Was that fun for you?' She was adorable: lively, lovely and worried about a small spot on her chin she was hoping Make-up had camouflaged.

I'd smuggled Allegra in. She was ten and having a rough time at school. Some nasty classmates had formed a spiteful gang and were tormenting her. Goldie spotted her and said she reminded her of her own daughter, Kate Hudson, just six years older. Goldie picked up on Allegra's wistful expression. She exuded such concern that Allegra stood close to her and told her the whole story. Goldie's team were waiting, but she had no intention of leaving a worried little girl until she'd filled her with hope for the future. She said a brutal bunch of schoolgirls put Kate through very similar misery. Bullying didn't feature on school protocols back in 1996. Children were left by teachers (and often parents) to sink or swim. They usually sank.

Goldie carefully explained to Allegra that happy people don't bully. Bullies only want to make others miserable because they're having a rough time. She told her to try to feel sorry for the bully. It didn't seem like psychobabble. I'd never heard anyone put it that way. Allegra brightened. Goldie hugged her. She checked her zit before she left. We said we couldn't see it.

We could, but why would you want to upset Goldie? We'll both adore her for ever.

Danny DeVito

Get in. Flog the product. Get out. That's the formula. Unless you're Danny DeVito. The man is a giant. Not literally, I know, but spiritually, intellectually, metaphysically. He arrived on the bed buzzing. He'd been to the theatre in Paris and seen a Russian play in French. Captivated, he couldn't stop talking about it. He didn't give a fig for promoting the film he was meant to sell. He wanted a conversation about something he was interested in. I'm not a Russian-drama maven, but I can have a reasonable bash at discussing Molière, Zola and Proust. Since Cambridge no one wanted to debate Diderot over Dumas with me. Danny DeVito was the exception. He is an interesting, interested man, alighting knowledgeably on topics he finds riveting. He thinks the memo saying, 'Sell, don't talk,' is bunkum.

Danny DeVito was the only guest – Harrelson never offered anyone a suck on his mango – to bring chocolates from the Bois de Boulogne in flamboyant boxes and distribute them to floor managers, make-up artists and best boys with gusto. He never got round to publicizing his film.

Wesley Snipes

Wesley made a flying visit to the UK. We couldn't afford to miss him, so I went in at the weekend for a pre-record. Pre-records suck. There's no adrenaline, no green ON AIR button, no oomph – unless you're pre-recording with Wesley. The man is a hoot. He riffs at an electrifying tempo. His thesis was that he'd like to stage a cops-and-robbers drama between my breasts. He liked the idea of climbing my twin peaks and pushing criminals to certain death in the crevice of my cleavage. Remember, in 1996 I thought I was happily married and there was no way I would risk my marriage and destroy my daughters' emotional stability for a cataclysmically carnal fling with one of the world's most devastatingly irresistible men, so I didn't. When my husband fucked off three years later, uppermost among my regrets was 'Damn! I bloody well should have shagged Wesley Snipes!'

Resisting orgies

The *Big Breakfast* bed was a cauldron of simmering sexuality. If I'd given way to my baser inclinations, I'd have enjoyed post-show nine a.m. orgies with KC and the entire Sunshine Band, Kool and his Gang's members, George Benson (who gave me tickets to his show in Nice. I gleefully attended), Cuba Gooding Junior, sexy as hell, and Charles Aznavour for whom we transported the bed to Paris. All were wildly attractive. All exuded USP, X factor, BDE and phenomenal pheromones. All were lying prone on the bed, just one unzipped fly away from a right rollicking. I was thirty-four, newly famous, ripe for seduction but wary of home-wrecking, especially mine. I politely declined.

Some no-thank-yous were easy. Others required more fortified moral fibre.

Jeff Goldblum

Independence Day and *Jurassic Park* hero Jeff Goldblum was quite simply sex on muscular, perfectly formed, exceptionally shapely, infinitely lengthy legs. Oh, my God, that man is pure paradise. One look at him and my ovaries somersaulted. He is funny and talks in puns, riddles and metaphors at a million mesmerizing miles a minute. He paints, plays jazz piano – he did both at once on the *BB* – and looks deeply into your eyes while composing an on-the-spot ode to your wrists as if, in you, he's finally unearthed the treasure he's been searching for all his life. I'm not sure I didn't fall in love with him on the spot. I've held a torch for him ever since.

Jeff Goldblum is the mensch against whom I measure all other men and find them wanting. He is brilliant, tender, vulnerable, unbelievably wealthy and so handsome he brings lubricity to the word 'moist'.

I think he liked me a little bit too. He invited me to a party in the West End that night. I went. I took my husband with me. I hadn't spent every second of my life waiting to get married to throw it away on a fling, even if it was with Jeff Goldblum, the unrequited love of my life. Jeff spotted me and waved me over. Am I overstating it to say his face fell as he saw my hand interlaced with my husband's? I don't want to big up the affair that never was because it never was. There's nothing more to say about Jeff Goldblum except 'He doth bestride the narrow world like a colossus'.

39

The worst bedfellows – Madonna and Miss Piggy

Third worst: Madonna

CHANNEL 4 WERE ITCHING TO capitalize on my forthcoming interview with Madonna. The PR department came up with *Clash of the Titans*-type promos: VANESSA MEETS MADONNA. MADONNA MEETS VANESSA. DON'T MISS *THE BIG BREAKFAST* TO FIND OUT WHO WILL BE THE LAST WOMAN STANDING. Madonna was what's known as 'a huge get'. In other words, getting her on your show is the biggest of all possible deals.

Here come the suits

When anything 'big' is on the horizon, the suits turn up. Instead of trusting programme-makers to rinse every ounce of potential out of the mega-guest, they descend from on high to put everyone on edge, dispatch vital crew for oat-milk lattes and throw the whole show off-kilter. Bosses arriving at the most delicate moment to chuck a spanner in the works is a regular occurrence. They stand in serried ranks throughout the very first show of every programme or series. They shouldn't. Head honchos gawping on the sidelines puts everyone off their A-game. If bosses do decide to pop in, it should be halfway through show seven when everyone knows what they're supposed to be doing.

After thirty-five years of peak boss at the wrong time, I still haven't plucked up the courage to tell any of them to naff off upstairs and stop raining on my parade. It takes balls to tell the chap who signs the pay cheques to take a hike.

Of course, the Madonna interview lured the suits out en masse. For once, *The Big Breakfast* had no choice but to join the throng. Madonna was too gigantic to entice on to our bed. On the morning of 19 December 1996 we were waiting in line with the rest of the world's TV, radio and print press for our allocated fourteen minutes at the only London junket held to publicize Alan Parker's film of Tim Rice and Andrew Lloyd Webber's musical *Evita*.

Channel 4 were determined my fourteen minutes would yield 24-carat nuggets of TV gold. Management I'd never set eyes on demanded to know my 'list of questions'. I don't do lists of questions. I'm always hoping to have a conversation, listen to my guests and respond to whatever they reply. I don't fire off one pre-arranged question after another. Management didn't want me to freestyle. I didn't want to commit to questions. The argy-bargy was courteous but not pretty. Make-up were tense. I was starting to sweat and shine. Hair were fraught. The venue was damp. In those pre-Brazilian-keratin-straightening treatment days, my sleek Veronica Lake blow-dry was frizzing up. Wardrobe were worried. Appliquéd beads were pinging off my ballgown. The looming line of bosses was freaking me out.

We were already having a horrible time when everything went into dismaying decline. Madonna was late. There was no explanation or apology. We waited. Japanese radio, Helsinki TV, the *Times of Canada*, *Paris Match* waited. CNN, the BBC, *Denmark*

Today and the *Guardian* waited. And waited. And waited. There were no announcements. Flights needed to be cancelled and rebooked. Childcare arrangements had to be extended indefinitely. Professionals who'd been outwardly cool about coming face to face with the biggest name on the planet started to come unstuck. Some had already missed live broadcasting deadlines. Some were scared they would. Print journalists, already under pressure to file copy before lunchtime, were chewing their pencils. Hundreds of us paced the floor. We were worried. If she was late, later, latest, would our precious time with her be cut?

Sporadic bulletins spewed forth. Fourteen minutes would be cut to eleven minutes. Management were *platzing* (fretting). Eleven minutes would be cut to nine minutes. Fire poured out of their nostrils. Nine minutes would be cut to seven. I swear I saw the biggest boss wipe away a tear. Channel 4 were billing 'Madonna and Vanessa' on the hour every hour as an 'appointment to view'. If Madonna didn't get her bleeding skates on, she'd be in and out like a flash of greased lightning. There'd be no time to blink. You'd already have missed it. Our 'get' was evaporating. The clock ticked. I didn't like this woman. We were there to publicize her film and she was treating us with casual contempt. She may be famous, I thought, but she's still just a human female – fucking with us all so carelessly. Allegra was ten, Saskia seven. I needed to be back to pick up my children. How dare Madonna play fast and loose with my family life?

Six hours later, I was pushed into the room for our – extremely – brief encounter. Madonna, who still hadn't explained or apologized for her tardiness, looked bored to tears, knackered and hostile. Her entourage sat on the floor tucking into takeaways. She didn't bother looking up or saying 'Hi' when I came in. I was fuming. On went the mic. The cameraman barked: 'Lean back, Vanessa. You're obscuring Madonna.'

'Good Lord,' I snapped. 'Obscuring Madonna would be the pinnacle of my career.'

Her head shot up.

'It's clear,' I said, 'you don't want to talk about *Evita*. So don't worry. I can talk about poetry, politics, sex, shopping.'

'Shopping!' said Madonna. We were off. At the start, she was monosyllabic but eventually I managed to bombard her with such unexpected questions, she couldn't help opening up. She said Lourdes, a couple of months old, was upstairs and she was exhausted breast-feeding her between interviews. She opened a present we'd bought and said the outfit – a tracksuit – was too boyish for her baby, who 'loved frilly dresses'. I told her Alan Parker said she had 'become' Evita. She seemed mildly gratified.

I asked if she was still into sex two months after having a baby. She said she hadn't gone off it. Finally, she smiled. She engaged. She complimented me on my bountiful bust. I said she'd looked fed up at the premiere in LA. She said: 'It was work', she was exhausted, and it had taken her an hour and a half just to get up the red carpet.

I asked if she'd become respectable. She said she always had been. Warmed up, she finally found the grace to sparkle. She didn't make it easy. She is the least 'people-pleasing' star I have met. Somehow, I managed to carve something worth watching out of six partially painful minutes. Channel 4 were so relieved, they sent me a giant bouquet. Madonna haunted my nightmares for years. Sometimes she still does.

Second worst: Miss Piggy

Miss Piggy was a prima donna and first-class pain in the arse. Mark Little was so incensed by her rudeness he hissed, 'Roast the pig!' at me on the stairs. Kermit and Gonzo were charming. Their puppeteers, led by world-famous Frank Oz, climbed inside the hollow *Big Breakfast* bed, smiled, schmoozed and dispensed sweetness and light. All was harmony until the green light went on and Miss Piggy swung into attack mode. Unwittingly, I'd offended her by slightly inclining my body towards Kermit and addressing him first. She was affronted, insulted and, however hard I tried, I couldn't win her over. She took aim at my technique, accused me of showing 'a bit too much leg, dear', told Kermit she was 'naked under here' and bewailed the fact that there were 'not enough roles for mature pigs in Hollywood, these days'. The three of them shouted over each other, indulged in unacceptably unsubtle double entendres, anarchy and insurrection. There were three mutinous Muppets versus one floundering Feltz. The victory – they turned our rendezvous into a battleground – was all theirs.

Bumping into Mark Little in the car park afterwards, I yelled: 'Bacon sandwiches!' He managed a wan smile.

40

The worst of all: Rolf Harris

Guess who you've got on the bed tomorrow, Vanessa . . . Rolf Harris. Rolf! If you'd promised Valerie Singleton and John Noakes with Petra and Shep I wouldn't have been more excited. Rolf was part of all our childhoods. I knew (and still do) every word of 'Two Little Boys' by heart, not to mention 'Jake The Peg' and 'Sun Arise'. We loved Rolf making funny breathy noises, while flipping a wobble board and splashing blobs of paint on a giant canvas, then stepping back to reveal that a tribal chieftain had magically appeared. Rolf was talented, gifted on the didgeridoo and a thoroughly good egg. The royal family were fans. He painted Her Majesty the late Queen.

When I started being invited to showbiz functions, good old Rolf often provided the cabaret. He'd fling out a few tired jokes and we'd laugh uproariously. Why? Because he was Rolf Harris, bastion of British TV, the bearded, smiling, paint-spattering Antipodean we'd grown up with. He didn't need to be funny to make us smile. He was Rolf Harris, for goodness' sake.

Rolf warbled his 1994 release 'Stairway To Heaven' in the *Big Breakfast* garden, then joined me on the bed. He oozed affability. In his slipstream waddled his smiling wife Alwen, recognizable by her beaded plaits. She jammed into the tiny bedroom alongside the floor manager and cameraman. Rolf and I boarded the bed. I was wearing one of my full-length, beaded ballgowns.

We were on air. He was jokey and full of beans. I heard a scratchy sound. It was the noise of the beads on my dress being crunched together as Rolf gathered up the fabric at my ankles and started pulling it further and further up my legs. I tried wriggling away from him. The bed was smaller than a real one. There was nowhere to go. I put the cushion I was leaning on between Rolf and me. It made no difference. He kept talking. His hand kept moving. I didn't know what to do. *The Big Breakfast* was light-hearted family viewing. Rolf Harris was Rolf actual Harris. A national treasure. A beloved institution. His adoring wife was three feet away. We were live on TV. His hand moved up my leg, past my knees towards my thighs. He was laughing, chatting animatedly as if everything was completely normal.

I knew I couldn't yell: 'Rolf Harris is assaulting me!' I didn't want to disturb our viewers. I didn't want to upset his wife. His hand curved round to my inner thigh. In a matter of seconds, he'd reach my pants. I couldn't risk it. There was absolutely no way I'd allow him to insert his groping fingers inside my underwear, much less inside me. No chance whatsoever. Yet, with the cameras on us, what could I do to stop him?

I did the unthinkable. Every TV presenter learns on day one that you never, repeat *never*, throw to an ad break unless the floor manager cues you. No matter how dull the guest or how rubbish the segment, you can't just decide you've had enough and fling to a commercial break. The ads have to be aligned. Precision is essential. The director decrees the moment is imminent. The floor manager makes countdown hand gestures, so you arrive at the break at the right fraction of a second. No presenter, not even Graham Norton or Jonathan Ross, rides roughshod over the process. That day, I had no choice. Just as Rolf's hand touched my knicker elastic, I leaped a foot in the air and spluttered: 'Oh, my goodness. Goodness me. Oh, my Lord. Quick! Let's go to a break. Now! Now! Now!' I disentangled his foraging fingers by jumping off the bed. In the break, I didn't say a word. Utterly unperturbed, Rolf chatted calmly to his wife. The cameraman shot me a sympathetic look. The ad break finished.

Rolf and I climbed back on the bed and made polite televisual conversation until our interview wound up. He didn't try touching me up in part two. When it was over I said goodbye politely and made for the door.

I didn't think of telling my Planet 24 bosses. I didn't make a complaint. I didn't want to ruin his marriage. I was thirty-four, married, a mother of two. I was fine, really, not defiled or damaged. Of course, and this is crucial, I didn't know Rolf was doing the same thing to other women. I didn't know he was doing it to underage girls, children. I didn't know his wife was often a hovering presence while he did it. I didn't know Rolf had ever behaved like that to any woman except me. I did know one thing: it wasn't my fault. I hadn't encouraged him, flirted or given him the green light to grope me. I told my husband about it. I mentioned it to a couple of my closest friends. I wondered why I found myself sobbing in the bath that night. After all, I was absolutely fine. What on earth did I have to cry about? I tried not to think about Rolf Harris after that and I managed pretty well, until, seventeen years later in 2013, there was a knock at my door.

Like the rest of the world, I discovered Rolf Harris had been arrested as part of a police investigation into sexual-abuse allegations, Operation Yewtree. I realized then I wasn't the only woman he'd assaulted. I decided not to go to the police with my story. After seventeen years, it was ancient history. I'd been an adult, not a minor. My life hadn't been blighted. I didn't think there was much to tell. Allegra and I were trying to interest my first grandbaby Zekey in solids, introducing him to mashed banana, when two police officers turned up. They had the footage of the Rolf Harris interview. Viewing the film, they said, my discomfiture was clearly visible. The moment Harris's hand reached the periphery of my pants was notable. The clip showed my mounting alarm and the moment I jumped in obvious panic. You could see me frantically throw to a break. How, the officers asked, would I feel about being a witness in the trial? I told them I didn't think my story would contribute much: nothing serious had happened. I wasn't keen. They asked me to reconsider because many of his victims, some of whom had been children when he assaulted them, were terrified Harris's reputation would override their testimonies. It would help them come forward if a well-known person publicly confirmed Harris had abused them too. Please, would I reconsider, agree to be called as a witness and, when they gave me the green light, give an interview telling my story? I agreed to both. I was never asked to appear in court.

I told (not sold) my story to the *Sunday Express*. Rolf Harris had just been jailed, convicted of twelve indecent assaults on four underage girls. One of the girls, his daughter's friend, had been assaulted mercilessly over a sixteen-year period. I was so vilely trolled that BBC colleagues Fiona Bruce and George Alagiah approached me (separately) in the corridor to say how sorry they were for the hideous onslaught. The messages, they told me, were along the lines of 'I thought I hated Rolf Harris for forcing sex on children. I hate the bastard nonce even more for wanting to shag that ugly fat bitch Vanessa Feltz.'

The *Guardian* ran an article about the vicious virtual attack. They were kind, but it wasn't necessary. I had only an eight-pound Nokia, no smartphone. I wasn't online. I was going about my business in blissful ignorance. If strangers hadn't given their condolences, I wouldn't have known a thing about the trolling.

More disturbing, though, was the sheer number of well-known female presenters who sought me out to confide Harris had done the same, or worse – sometimes far worse – to them. I won't reveal their names, but if you think about the roster of female talent on TV at the time, there's a very strong chance any name that pops into your head will be among them. I was told women had been warned never to go in a lift alone with Rolf on his show *Animal Hospital*. I suppose our collective silence stemmed from fear of being hated for toppling an idol and fear of being branded a trouble-maker. #MeToo didn't gather momentum till the exposure of numerous sexual-abuse charges against powerful film producer Harvey Weinstein four years later in October 2017.

41

Coming up, two great Birds-Eyedeas

By 1996, I was presenting *Vanessa* on ITV, *The Big Breakfast* on Channel 4, *Value for Money* – 'the show in the know when it comes to shopping' – on BBC1 every Thursday night at 7.30 p.m. in the summer months and *Vanessa's Day With* . . . ? (Boy George, Paul Young, Edwina Currie, Max Clifford – who was later jailed for rape, Karren, now Baroness, Brady, et al.) on Channel 5, plus a Sunday-morning radio show on the new Talk Radio and writing a weekly column in the *Daily Mirror*. Ubiquitous? I was inescapable.

I even landed a delightfully lucrative voiceover gig for Birds Eye. You wouldn't believe the hordes of advertising executives involved in recording a big-budget voiceover. It looks like a crowd scene from *Ben-Hur*. The client, Cap'n Birdseye himself, arrived mob-handed. At least twenty-five Birds Eye bigwigs milled about. The ad agency sends battalions of account handlers, plus juniors to wait upon their every whim. There are teams of 'creatives', producers, dozens of sound engineers, PR people, directors . . .

Unprepared for the throng, I walked into the agency solo. Someone handed me a script. This is what it said: 'COMING UP, TWO GREAT BIRDS EYE-DEAS.' I read it, once, twice, three times. I waited. A flurry of producers descended. 'Have you familiarized yourself with the script, Vanessa?'

'Are you across it?'

'Have you mastered it?'

It was six words. I felt reasonably confident I'd grasped the thrust. They ushered me to the sound booth, poured me water and handed me the headphones. 'Darling, are you ready? Do you feel up to trying a take?'

'Yes, thank you. I'm fine and raring to go.'

'All right, Vanessa, we're going for upbeat, positive enthusiasm, tempered with approachability. We want aspirational meets down-to-earth. Keep the voice light, but earthy. Persuasive, but not too pushy. Authoritative but not bossy. Hot but chilled. Got it? OK, VANESSA, BIRDS EYE-DEAS – TAKE ONE!'

With every ounce of positive earthy hot/cold lightness I could muster, I said: 'Coming up, two great Birds Eye-deas.'

There was pandemonium. The creatives converged on the execs. Botoxed brows were figuratively furrowed. People scribbled stuff in notebooks. Execs whispered animatedly to creatives. It took for ever. Alone in the soundproof booth, I could only wonder what on earth they were saying.

Red in the face, the producer gabbled: 'Vanessa. Darling. That was wonderful. Tremendous. It's in the can. We're ecstatic. Just brilliant. Brilliant. Genius. Indulge us, darling. We'll go for another. This time, ramp up the liveliness, but don't make it hurried. Deliver warmth, darling, lots of warmth, but not gooeyness. Keep on the right side of gooey. Pick up the pace just a scintilla, but don't rush it. Take your time, but swiftly. Don't linger. Drive home the message but keep it gentle, gentle, gentle. We want sweetness and light, but not too much of either. Got it?'

'Yes,' I said.

'OK. VANESSA, BIRDS EYE-DEAS – TAKE TWO!'

'Coming up, two great Birds Eye-deas,' I said. To my untrained ear take two sounded identical in every syllable to take one.

Frankly, I couldn't understand the directions, and even if I could, I didn't have the faintest idea how to ramp up, but slow down. I was flummoxed, but the fee was far too fat to admit it. I kept smiling and nodding and saying, 'Coming up, two great Birds Eye-deas,' in exactly the same way for six hours straight. Take thirty-seven was the clincher. The client went wild. The agency was ecstatic. I was so relieved to exit the booth I forgot to take my month's supply of free fish fingers.

Whenever I threw to a commercial break during a show, I could hear my own voice proclaim: 'Coming up, two great Birds Eye-deas.' I sounded hungry, but not ravenous, warm but not slushy, and like a close personal pal of Cap'n Birdseye. Every time the ad was on telly, my coffers went 'Ker-ching!' It wasn't a one-off one-price buy-out. It was a more-the-merrier downright splendid earner – cherry-on-the-cake time. They booked me to announce tidings of comfort and joy: 'Coming up, two great Birds Eye-deas for Christmas,' and 'Coming up, two great Birds Eye-deas for Easter.' I could see the job running right round the calendar. 'Coming up, two great Birds Eye-deas for Valentine's/Chanukah/Ramadan/weddings/funerals/total eclipses/circumcisions.' Unfortunately, the job stopped at Easter, and so did the cheques. They were sumptuous while they lasted.

42

What it's really like to be famous

I WAS FAMOUS. SHOP ASSISTANTS stared, blushed and gave me the wrong change. People passed me in the street, did a double-take and walked backwards to pass me again just to check. Autographs were still a thing. I'd practised mine on my rough book at school. It was big and swirly with kisses. I was asked to sign someone's leek and potato pie. I was given a felt-tip to inscribe the flesh just above a chap's pubic bone. I was grown-up, grounded, not an impressionable teenager itching to drive a Jensen Interceptor into a swimming pool, develop a dangerous heroin habit and swipe Mick Jagger from Jerry Hall.

Let's be frank, though, going from anonymous to famous is a bit destabilizing, chiefly because people who don't know you act as if you're a big deal. For years no one, including your relatives, gives a damn if you show up or not. Suddenly senior executives are pacing the pavement, ringing the car company, beseeching them for your precise ETA. When you arrive, their relief is almost palpable. They are chuffed. They carry your bag. They'd carry you if you let them. 'Vanessa can walk but thank God she doesn't have to.' They rush to bring you coffee. If your coffee order is complicated, they like it more. They seem to want to dash about doing impossibly demanding stuff on your behalf. If you bring a sandwich from home and drink tap water, they're disappointed. I wondered why. On reflection, I think it's because they want to be able to tell a good story about you afterwards. If you're quiet, polite and undemanding, it's boring. They want you to ask them to take their coat off their back and give it to you. They want you to demand ketamine, a colossal dildo, small white kittens and egg fried rice without the egg, so they can work it into a fabulous pub-worthy anecdote. If you're a people-pleaser, the temptation to act up and give them what they want is immense.

People – including those you think might know better – act differently towards you when you're famous. Some laugh exaggeratedly at your jokes, fawn all over you, pay you oleaginous compliments and crawl so tenaciously in the direction of your anal orifice you want to disinfect yourself (and them). Some assume you must be standing on a pedestal loving yourself and do their utmost to topple you, like the smashed-up statue of Saddam Hussein.

You're in the dentist's waiting room happily reading a copy of *Woman's Realm*. They interrupt accusingly: 'I've never heard of you, but my wife says you're famous.'

What are you supposed to reply? If you say, 'I'm Vanessa Feltz. I present a TV show,' they make a giant palaver out of never having heard of or seen it. If you say, 'If you haven't heard of me, I'm not famous to you, so don't worry about it,' they get offended and think you're a Billy Big Bollocks.

There are two premium hardy-perennial humdingers.

1 'I always thought you looked awful on telly, but actually you're really not that bad-looking in real life.' How the heck do you respond to that thumping great back-handed compliment? Dame Maureen Lipman tells them straight: 'So you're saying when seven million people were watching me on prime-time

telly, I looked hideous, but standing in front of you in the post office in East Finchley on a wet Wednesday, I look just about passable. How rude did you mean to be?'

2   Always uttered in tones of outraged exasperation: 'I bet you don't remember me!'

What a stinking opening gambit! It puts you on the back foot immediately. Usually I smile, act as if I'm ferreting around in my brain for a name and say: 'Ah! I knew I recognized you. Remind me.' The answers would be hilarious if they weren't delivered with such deadpan sincerity and if I hadn't run the you-don't-remember-me gauntlet for thirty solid years.

'Susan Grimthorpe. You know. You stood under my umbrella at Piccadilly Circus when you came out of the hairdresser in 1994.'

'Dilys Pilchard. I asked you for directions when you switched on the Norwich Christmas lights in 1997. Don't you remember?'

It would be unforgivably churlish to reply: 'What? Are you serious? Of course I don't remember you.' So you don't. Sometimes I apologize sincerely to total strangers several times a day for failing to recollect the moment they handed me a copy of the *Evening Standard* in early 2013 or lent me a comb in the loo at Monty's nightclub in 1998.

The all-purpose grin

Some famous people don't care if you like them or not. When I filmed *Vanessa's Day with Boy George* in 1995 he'd ask intrusive autograph hunters, 'What's the last album of mine you bought?' If the answer was *Colour By Numbers*, the one with 'Karma Chameleon', twelve years earlier, he'd tell them in no uncertain terms to do one. When I ran into my *Big Breakfast* colleague Mark Little on holiday in Malta, we combined families for dinner. When fans saw us in the restaurant and clustered over for an autograph, Mark roared: 'Oy! We're trying to have a holiday here! Fuck off!' My girls, seven and ten, were enraptured with his X-rated expletives and fascinated to see an adult being so unashamedly rude.

My style is very different. Don't ask me why, but I'm so anxious not to be thought supercilious or up myself that I keep an all-purpose grin plastered all over my features whenever I'm out and about in public. My face aches from smiling. I have stretch marks round my mouth. I just don't like the thought of someone saying to her sister-in-law, 'I saw that Vanessa Feltz at the crematorium today. What a surly sour-faced cow she is.' The most delightful thing about flying off to somewhere no one's heard of you is finally being able to relax into a resting bitch face.

43

Fun facts about being famous

1 Polite household names introduce themselves by name. I was in a lift with Dawn French. She said charmingly: 'Hello, I'm Dawn French.' Of course she knew I knew she was Dawn French. Everyone knows Dawn French. She's just beautifully brought up and wouldn't want to presume. Naturally, every human on the planet replies: 'I KNOW you're Dawn French, Dawn French.' That won't stop her doing it. Nice famous people all do it. Sir Elton John isn't necessarily known for being a gentleman with the sweetest of tempers, but he does it too. 'Hi, Vanessa, I'm Elton.' I wanted to kiss him for his excellent manners. I managed not to make lip contact, but I massively felt like it. Sir Rod Stewart does it. He and his thoroughly decent, kind-hearted, warm-spirited wife Penny are both unassuming. 'Hi, he's Rod, I'm Penny. This is my dad. This is my school-friend. Great to see you, Vanessa.' The Stewarts were supporting a mate at the Breast Cancer Care Ball at the Grosvenor House Hotel in London's Park Lane. They were swamped by fans but much too polite to show signs of impatience. In my experience people who think they're too famous to bother introducing themselves are usually reality TV 'stars' who have spent the summer shagging other irresistibly attractive youth on the small screen when their grandmas could be watching.

2. Famous people give other famous people a special famous-person grin and nod. I've had the we're-both-famous grin and nod from Bill Nighy, Brenda Blethyn, Max George from The Wanted, Jeremy Paxman, Noddy Holder, Sir Magdy Yacoub and, I kid you not, in 1996 on a Tuesday afternoon in Beauchamp Place, Knightsbridge, the late Princess Diana. My children think it's hilarious. 'Oh, my God! That was Billie Piper giving you the we're-both-famous nod. Lord, Mum, how do you know her?' The answer is: 'I don't.' I'm not sure I've ever actually been introduced to Billie Piper. I've never even had the pleasure of working with her. It's true, though, that when we both happen to be having lunch at gorgeous Primrose Hill Greek restaurant Lemonia, we do give each other the we're-both-famous grin and nod. It would be rude not to. She was married to Chris Evans. I've worked with him – and gone in deep for our special Friday Hug – on *The Big Breakfast* and for years at Radio 2. She's been hounded by the tabloids. So have I. She's been mercilessly trolled. So have I. She's had two horribly public divorces. I've had one horrific public divorce and a hideous public break-up. Of course we give each other the grin and nod. It's shorthand for 'I don't know you but I get you. I'm not you but we've walked a few wobbly steps in each other's Jimmy Choos.'

3. Landing at Heathrow – or any lesser British airport – is much cosier if you're famous. (I've only flown on a private jet once, to Milan for *CelebAir* in 2008. It was where I met my late lamented friend the 'Holy Focaccia, Vaginika Semen, Chinese Girls with Herpes' maverick, David Gest. He asked Amy Lamé, one of the celebs competing for the title 'Best Flight Attendant', to floss his teeth for him. She did, but Lisa Maffia won the series. The experience was wild. Imagine being paid to be waited on hand and foot while being private-jetted to Italy for a multi-cocktail lunch and back. It was one of my best gigs of all time.)

4 Disembarking any flight, even the budget kind, is a jamboree if you're famous. The passport queue is like a personal party. Everyone wants to know where you got your tan. Did you have a good time? What do you think of the gossip about Michael Barrymore? The Passport Control person takes a double-look at your pic and says: 'Vanessa! Lovely to have you back. Welcome home.' For me, it's one of the best things about being famous. You feel warm inside. You feel as if you've come back to your beloved country where people love you. You feel, for a few minutes, as if you matter. It's a wonderfully nurturing sensation. You'd love it too. Sometimes it only lasts till the taxi queue where someone stage-whispers: 'Over there. Isn't it that gobby fat bird off the telly?' Still, it's magnificent while it lasts.

5 Famous people refer to non-famous people as 'civilians'. Allegedly Liz Hurley coined the phrase. I was taken to dinner at eye-wateringly expensive London restaurant Harry's Bar in 2023. Two people exuding other-worldly glamour were dining. She was eight feet tall, willow thin and draped in bronze sequins, her flawlessly tanned back exposed and gleaming. He was peacock-bright, groomed to a satin finish. They flashed luminous piano-key smiles at me. It was the famous-person-grin-and-nod. I gave them my best Cheshire cat impersonation right back. It took me a while to work out who on God's earth they were. Only Liz Hurley and David Furnish. I've never met either. I've interviewed David over the radio airwaves but never seen the burnished reality. I was on a dispiritingly wretched first and only date with a shrivelled but loaded antique, who explained in detail why he usually orders the Dover sole but doesn't like the way they leave the skin on the bottom at Harry's till I wanted to choke him with the porcelain finger bowl. I itched to wave a sandwich board at Hurley and Furnish, saying: 'I AM NOT WITH HIM. I DON'T

Grandma Babs and Poppa Mo.

Poppa Mo at his most Clark Gable.

Grandma Sybil, Poppa Willie and my mother Valerie.

Grandma Sybil with my mother on her wedding day.

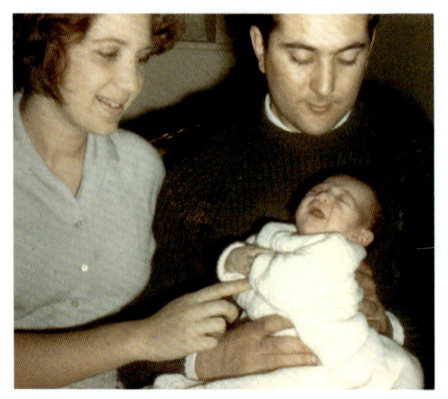

Newborn with my parents, Valerie and Norman.

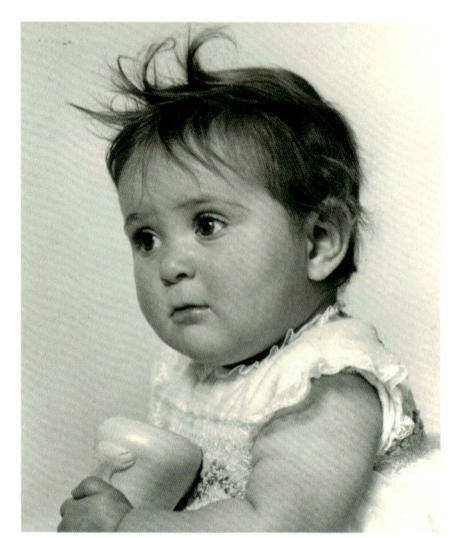

Aged 2, with my father Norman.

Aged 18 months.

Aged 2, with both my grandmas, the Yiddish chorus.

On my third birthday.

As a bridesmaid, aged 8.

Developing, aged 9.

Developed, aged 12.

Trinity College May Ball.

With my mother on my wedding day, aged 23.

With the girls in yom tov outfits, 1991.

My first stand-up routine, being introduced by Peter Moss, 1991.

Early days presenting Jewish London *on GLR, 1992.*

In NYC, 1994. *Learning from Phil Donahue, 1994.* *Co-hosting* The Jerry Springer Show, *1994.*

The stretch limo sent to Finchley to whisk me off to Hollywood, 1997.

Launching Vanessa ...

with Allegra and Saskia, 1994.

My first studio audience.

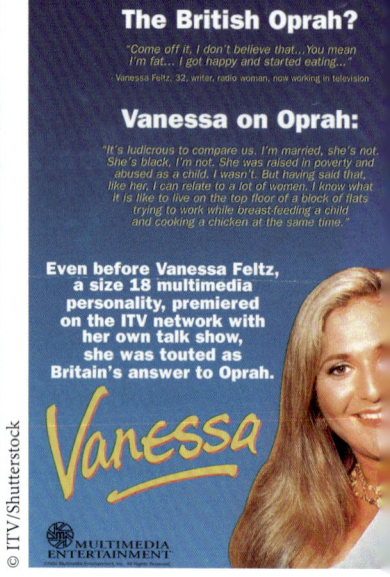

International syndication poster for Vanessa.

My favourite house in Finchley, which I reluctantly sold to upgrade to the rock-star mansion.

Allegra and Saskia at Allegra's batmitzvah, 1998.

December 1999. Down, but ... not out – bouncing back in the year 2000 ...

and at the MOBO Awards, 2002.

Me in a shopping trolley with lard equating to the weight I lost by working out.

My fitness video and nationwide tour.

Me with Boy George at his fortieth birthday party, 2001.

Revelling in my new look for John Swannell, 2000.

In Celebrity Big Brother, *2001, Saskia photobombing.*

Oh yes I did! Aladdin, *Woking, 2001.*

Celebrity Fit Club, 2004.

Me at a masked ball with Anthea Turner, Esther Rantzen and Gloria Hunniford, 1999.

Me and Peppa Pig in 2016 – which one's which?

Me at the Sony Awards, 2007.

© Glenn Copus/ANL/Shutterstock

With the London Mayoral candidates at BBC Radio London in 2008.

Winning the Sony Award in 2009.

My first day at BBC Radio 2 in 2011.

With Jeremy Vine at Radio 2 Live in Hyde Park in 2013.

The telly years – the best: with Joan Rivers on the Big Breakfast bed …

and the worst: with Miss Piggy, 1996.

Channelling Cher on the Titanic for Comic Relief, 2013,

and Meryl in Mamma Mia! for Children in Need, 2012.

Starting Strictly, 2013.

With my friends Alison Hammond and Holly Willoughby on This Morning.

Arriving via motorbike to greet my biggest fan, Jack Whitehall.

Fame: having fun with all the best people, I am with Lionel Richie, 2001 ... and Chris Evans, 2012.

Marking forty years of shopping at Brent Cross, 2016.

The then Prince Charles remarking 'Oh Vanessa, Vanessa, Vanessa', 1994.

Turning sixty on This Morning, *2022.*

Out every night, 2024.

With some of my favourite people, including Jonathan Ross, 2001.

Piers Morgan, 2005.

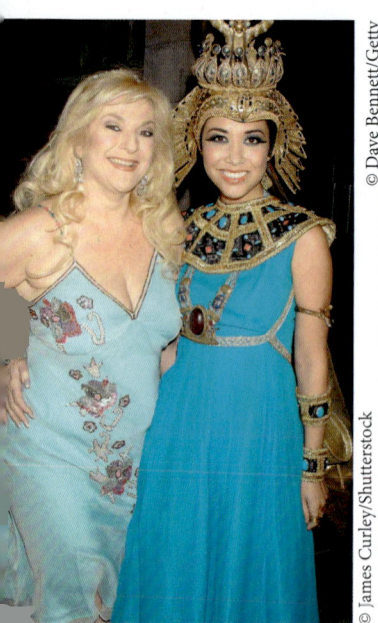

Myleene Klass at her thirtieth birthday party, 2008.

Lisa Snowdon at her book launch, 2024.

Elizabeth Day at her podcast launch, 2024.

With my gorgeous girls, Allegra and Saskia, at the Breast Cancer Care Fashion Show, 2011.

At home, 2005.

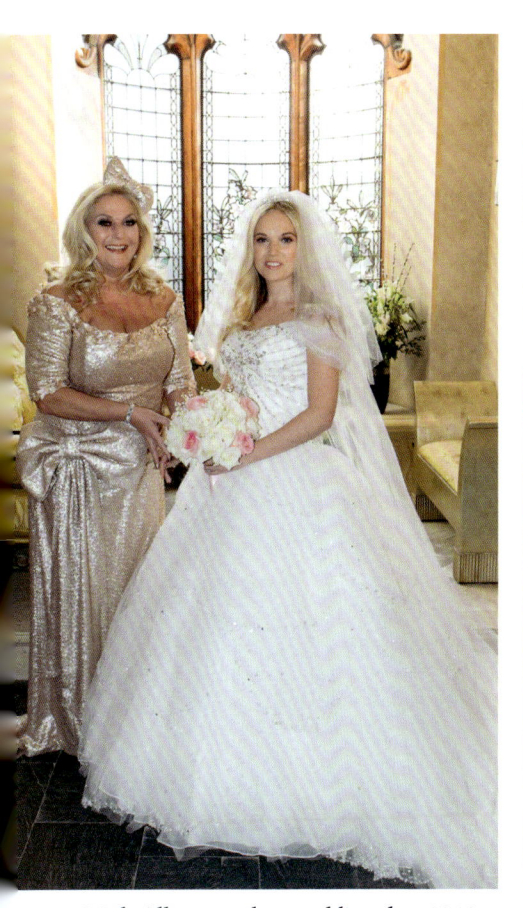

With Allegra on her wedding day, 2013.

With Saskia on her wedding day, 2016.

With my gorgeous grandbabies at the Imperial Hotel, Cork, 2023.

In our Christmas jammies, 2023.

My fabulous family, 2024.

At Ballynamona Beach, East Cork, 2024.

EVEN KNOW HIM. HE WAS AN INTRODUCTION FROM A NEIGHBOUR I NOW REALIZE DOESN'T LIKE ME VERY MUCH.'

6   Being recognized wherever you go is hell, especially if you're with someone you're not supposed to be with, frolicking in an unsuitable position, and Tom, Dick and Harry all have smartphones welded to their weirdly shaped thumbs. Being (in)famous means there's a buoyant market in sales of pictures of you doing something you shouldn't be doing. Their value increases the more offensive whatever you are up to would be to *Daily Mail* readers. If you are fat with a mouthful of doughnut, that sells. If you are fat with a doughnut in one hand and the penis of someone else's husband in the other, that sells even better. If you are weeping as you exit an STD clinic, ditto. I have seriously contemplated purchasing a burka just to be able to go about my nefarious business in private. I didn't do it because (a) do you remember the pictures of Nigella Lawson trying to hide her lumpy bits on an Australian beach swathed head to toe in a black burkini? She said her then husband Charles Saatchi didn't like her with a tan but, frankly, the look didn't do her any favours. And (b) there's always the prospect of the person who sold you the burka ringing the press and dobbing you in. When it happens, it's quick – 11.17 a.m. exit excellent shop with carrier bag containing two burkas, one to wear, one in the wash. 11.27 a.m. *MailOnline*: 'Has she converted to Islam? Why has Vanessa Feltz snapped up two burkas? Friends say she's seen the light.'

   Being recognized everywhere is almost as bad as

7   NOT BEING RECOGNIZED.
    You're finally abroad. Phew! What a relief. No one has the faintest idea who you are. You're elated. You're liberated. You can look sullen and miserable. You can air your cellulite. You can

give the fondue a double dip. You can do something downright disgusting with the deckchair attendant. You can score oceans of miaow-miaow. Only, the waiter is charmingly efficient; the deckchair attendant is a happily married lesbian. You don't know if miaow-miaow is a pill or a powder and you'd rather have a Cornetto anyway. Of course, the moment you can indulge in a smorgasbord of heinous, reprehensible activities without fear of detection, you lose all enthusiasm for sin. Your face feels floppy and pointless without the famous-person grin. You look hopefully around for just one British tourist to interrupt your peace with an annoying autograph request. There isn't one. You can read the excruciating book you've been carrying around for months. You will not be disturbed because no one cares you're there. You hate it. You hate yourself for hating it. You're thinking the unthinkable. I didn't get famous for peace and quiet. What the fuck is the point of being famous if no one recognizes me? You hope the suspicious-looking cove on a jet ski is a pap. He isn't. You ring home to see if anyone's talking about you. No one has noticed you're gone. You text pals: 'So rested. Heaven on earth. What a magical escape.' You cannot wait to be pestered by bored holidaymakers on the easyJet home.

8 You get in to places you're not invited to even when you don't want to. Matthew Wright requested the pleasure of my company at his birthday party at a club in Leicester Square. The doorman practically threw his arms around me. 'Vanessa! Lovely to see you. They're on the fourteenth floor. Take this lift straight up.' The buffet was sublime. I didn't hold back. I'd demolished three chicken legs and was knee-deep in the beetroot salad when I began to wonder where the Birthday Boy was? And his friends? And his family? And the team from *The Wright Stuff*? They were nowhere to be seen. I'd only inadvertently crashed and pillaged

the BT summer party. I offered to pay for the tuck I'd munched in error. BT staff were charming. A cavalcade of selfies ensued. I tracked Matthew down in the end.

Wish You Were Here . . . ?

Wish You Were Here . . . ? was a TV institution. It began in 1974 when 'foreign' holidays were an exotic fantasy for most viewers. Broadcasting goddess Judith Chalmers, tanned to a rich mahogany, would go on a package break to Lanzarote and explain how you cope with the luggage carousel and the all-you-can-eat buffet. She'd talk to Bob and Vera from Scunthorpe, roasting to a crisp, chargrilled noses peeling, and click castanets with them at the flamenco-dancing display. The show went out weekly on ITV until 2003 when every single person in the UK had been to Benidorm and survived intact. Lucky celebrities were chosen to be filmed having a ball on a free holiday.

In 1995, that celebrity was me and I was allowed to take my husband and children along. We were sent to the luxurious, million-miles-out-of-our-price-range Forte Village in Sardinia. Fresh gardenias were placed in our en-suite loo daily. I mean it. That is how the Other Half pees. They wee on exotic flora. The girls had a fine time. I forged a sustaining friendship with my adorable Mancunian chum Sara, and we all tore around on electric scooters loving and living the high life. *Wish You Were Here . . . ?* sent me (and the Good Doctor) to Monte Carlo. We flew there by helicopter from Nice airport, drove a Formula 1 car round the circuit and saw Paul Anka ('Put Your Head On My Shoulder') in cabaret.

In 1998 *Wish You Were Here . . . ?* sent us skiing in Aspen. I couldn't ski. The Good Doctor could. I didn't want to leave the girls for too long. He flew out ahead (on my gig) and took full advantage of the opportunity. Should I have seen something sinister in his eagerness to savour my job without me there to share it? Maybe I should have, but I didn't. I was just happy he was happy.

44

Co-host The Jerry Springer Show *in Chicago*

MULTIMEDIA IN NEW YORK WERE worried. They didn't fancy their British talent (me) cropping up on their rival US talk show *Maury Povich*. They had a problem. They couldn't stop me. My contract was for a British show. When it was drawn up, they thought I was a parochial British island-dweller who wouldn't need international confining. Wrong! Maury loved me. I loved popping across the pond first-class to luxuriate in his love. What was not to adore? Multimedia dreamed up a plan designed to stop me cuddling in Maury's warm embrace ever again. Why didn't they fly me over to Chicago to co-host their now-roaring success, *The Jerry Springer Show*?

Why indeed? Jerry Springer was unknown in the UK in 1994, deemed too sordidly salacious for a sheltered British audience. When I told chums I was off to co-host with Jerry they shrugged. 'Who?' Obviously, guilt prevented my showing a flicker of excitement at home. I couldn't just say exultantly to my husband: 'Wow! I've never been to Chicago. Cor! This is well brilliant! Yippee!' I had to feign indifference, fill the freezer with food for every single meal that would be eaten in my absence, bribe the au pair not to mutiny, and wear sackcloth and ashes while flagellating my bum with birch twigs. In 1994 being a working mum was only permissible as a penance.

Talent-handling is vital but no boss ever bothers

I landed at glacial pre-Christmas Chicago O'Hare. Pitching up at Springer headquarters, it was clear there had been a monumental cock-up. No one had bothered to 'talent-handle' Jerry. Talent-handling is a vital showbiz skill possessed by almost nobody. Keeping the talent happy, loved and loyal is essential to any station's success, yet no controller ever bothers to appoint anyone with the aptitude or savvy to make it happen.

Think about it. Talent is always defecting. Richard and Judy decamped from ITV to Channel 4. Graham Norton quit Channel 4 for the BBC. I . . . Spoiler alert! Alas, details of my own ill-fated exodus are coming up all too soon. Even Morecambe and Wise deserted the BBC for ITV straight after twenty-eight million fans giggled their way through their 1977 Christmas special. Why all the chopping and changing?

Performers perform to feel loved – cash is fab, but it isn't everything – and no one in a suit seems to have the smarts to realize it. It doesn't take much to make talent bask in appreciation, just someone in management to pop in once a month or so and say: 'Magnificent job, old boy. Highest ever ratings. We couldn't love you more.' Talent thrives on a bunch of flowers on its birthday, a bottle of decent red at Christmas, an appreciative outing to the races, a text message every so often: 'Terrific show today. Board room buzzing.' It never happens. Without regular schmoozing, back-slapping and expressions of gratitude, talent feels it is 'blushing unseen', withers and – before it dies – moves on to another station it hopes will buy it a pair of cufflinks and tell it it's iconic.

No one had bothered to pre-schmooze Jerry. All they needed to whisper in his shell-like was 'This British broad's coming in. The company's trying to keep her sweet and stop her doing *Maury*. It's just a gesture. It means nothing. She's new to the business. Humour us. She's gonna co-host, but only for one show, Jerry. Then she boards a British Airways and we never see her again.' It would have taken them less than a minute to woo Jerry and make sure he was happily on board. They didn't. Jerry was fuming and on the defensive. He thought I was making a play for his job and management were trying to inflict me on him like an incurable curse. He was livid. He refused to make eye contact or communicate in any way, save savagely swinging a precious baseball bat – I think it belonged to Babe Ruth or something – in a hostile manner during briefings.

Our show was called: 'I Had Sex With My Aunt'. I say 'our' but every day Jerry hacked away at my part in the proceedings. First, he decreed I couldn't stand where the producer had placed me, next to him on the perimeter of the set. Then he decided I couldn't stand in the well of the set near the audience either. Then he decreed I had to stand with the incestuous family on the stage. Each time Jerry laid down the law, the production team bowed to his almighty will. Then Jerry whittled down my bits on autocue. I was meant to introduce part two. Jerry put his own name on the intro. I was meant to throw to a break in part three. Jerry insisted only he could do it. I didn't know him well enough to say: 'Relax, mate. I couldn't care less about this. I have a husband, two children and a life I love in the UK. I don't want your show and they're not going to offer it to me anyway.' So, I politely accepted whatever paltry dregs Jerry permitted.

Twenty years on, the show is a blur. In fact, it was a blur at the time. The family was surly and snarling. The aunt was anything but a sexy siren. Jerry did his usual heavenwards eye-roll, implying none of the mayhem had anything to do with him. Fist-fights erupted and had to be broken up by the in-house security team. When I spoke up, sounding more British than Her late Majesty, the audience didn't have a clue who I was or what I was there for. By the time I'd had to jump off the stage to avoid being punched in the head, taken refuge in the wings and listened in silence to Jerry imparting celestial thoughts, I couldn't wait to get the hell home.

A couple of years later Jerry, now a household name in Britain, flew over to guest on some chat shows. Our eyes met. He greeted me like one of his oldest and dearest pals. He gave me a bracelet from Tiffany for old time's sake. In his mind, he'd evidently rewritten my sojourn on his show as a meeting of kindred spirits. It would have been ungracious to shatter our *entente cordiale* by telling the tale from my perspective. Jerry was an enchanting dinner companion and an adornment to my life. He died in 2023, lavishly lamented.

45

LA calling

EVERY FRIDAY NIGHT, JUST AS I was seasoning the soup, the phone rang. Every time I answered some American stranger said he wanted to fly me out to LA. Every time I barked, 'You must have the wrong number!' put down the receiver and got on with slicing the courgettes. I thought the fellow was a crackpot stalker weirdo. I had no idea how he'd got hold of my number and I didn't waste time worrying about it. There were children to bath.

On his fifth or sixth call the bloke managed to blurt out enough information to stop me hanging up. He was an LA-based programme-maker/talent-spotter; he'd seen my showreel. (I'd forgotten I'd even made a showreel. He might have meant the film Malcolm knocked up to show the Americans I could listen. I'm not sure.) Anyway, he liked what he saw, and he thought I'd be perfect to present a US talk show, based in Hollywood, to fill what was about to be a space left by unsuccessful programme *The Mommies*. He was so keen to meet me, introduce me to the appropriate American programme-makers and engineer a deal, he wanted to pay to fly me over, put me up at the Beverly Hills Hotel and see if he could shoehorn me into a US career. He seemed to be legitimate. I talked it through with the Good Doctor, phoned in sick to *The Big Breakfast* and – I'm not kidding – got into the stretch limousine he sent to pick me up in Finchley, N3. 'Bye, kids. See you around. I'm off to Hollywood.'

I was only gone for two nights. He schlepped me round mega-agents CAA. I glad-handed ABC bosses who asked me the obscene-sounding question: 'Have you been beveraged?' Taken to lunch at the Disney executives' private restaurant the Rotunda, I bumped straight into my *Big Breakfast* boss Charlie Parsons. 'Vanessa? Heard you were ill! So glad you've made such a quick recovery', he said, without missing a beat.

The Americans were keen. How would I feel about moving out to LA and making a coast-to-coast talk show? Dollars were discussed. Silly money. I lay on the bed in my suite at the Beverly Hills Hotel. I'd been taken to the Hollywood Walk of Fame and (inexplicably) introduced to veteran US TV star Dick Clark. My head was swimming with dollar signs and brimming with possibilities. I flicked on the TV. Public-service advertising boomed: 'KIDS, PUT DOWN YOUR GUNS.' I rang my husband, bubbling with the news. 'Darling, imagine the adventure. We could move out here for a while and see what happens. It would be amazing to try living in Hollywood. I can't quite believe this is happening?'

The Good Doctor was firm. 'Of course we're not moving to Hollywood. Don't be stupid. The children are settled at school. We're just about to complete on our new house. I'm a surgeon. I can't just jet to America and go into practice. I'd have to requalify and sit US exams. I have no intention of doing that. It's out of the question. Don't be absurd. And, anyway, *how much more famous do you need to be?*'

And that was it. Discussion over. Prospect over. Possibilities nuked. I've never forgotten his rhetorical question: *How much more famous do you need to be?* Gentle Reader, my enthusiasm for the idea had nothing at all to do with wanting to be more famous. That wasn't the USP at all. The girls were nine and six. I thought it was the perfect time to take a chance on an unlooked-for adventure. I fancied giving the job a whirl, just to see what might happen. Fame wasn't the spur. I wanted to try it.

The TV message reverberated in my brain: 'KIDS, PUT DOWN YOUR GUNS.' My husband's list of objections overruled my frivolous wish to up sticks and have a bash. How dare I drag my girls out of safe North-west London into the 'hood? How dare I contemplate disrupting my daughters' education for a silly TV show? Remember, I was raised to be a Jewish geisha. Derailing my husband the doctor's career, even temporarily, for the sake of my own, was unconscionable. Force a surgeon to requalify? Don't be ridiculous. How dare I, even for several million dollars that would have set us and the girls up for life, think of putting my career – if that's what this sure-to-be-short-lived flirtation with show business really was – before the good of my own family? I could hear my dead mother's voice: 'Vanessa, forget about it.' I flew home and never mentioned the offer again.

The Americans took my silence as a signal to stump up more money. They rang my solicitor Graham Shear daily, yelling numbers at him. He reluctantly told them: 'Unfortunately, no, thank you really means no, thank you. She's not playing games.'

When my husband walked out in 1999, I wondered why he hadn't at least let me try. There'd have been more in the kitty to divvy up. I would probably have failed, but I might have succeeded. Knowing he didn't love me and wasn't going to stick around, why did he make sure the US proposal was dead in the water? Twenty-six years on, I still can't work it out.

46

Famous people get to do weird stuff

Here's my shortlist of unexpected side effects of fame.

1. I ate a live locust on Darren Day's Saturday-night show *You Bet!*. It was scratchy, like swallowing a dry twig. I smiled for the cameras and coughed for weeks afterwards trying to dislodge bits of locust from my throat.

2. My girls met the Spice Girls. If you were alive at the time, you'll know that was a more significant guarantee of parental Brownie points than a face-to-face with the Almighty himself.

3. I was Gotcha'd by Noel Edmonds on his *House Party*. Embarrassment at the hands of Edmonds was a compulsory celebrity rite of passage from 1991 until 1999. If you hadn't been Gotcha'd you weren't bona fide. If you had, eighteen million viewers gloried in your humiliation and would never forget about it or let you forget about it in a vicious loop of shame. I dreaded a Gotcha because I'm a swearer, addicted to Anglo-Saxon expletives. To this day if I take a taxi and have a chat with the driver, s/he'll turn round in amazement and say: 'Bloody hell, Vanessa! I never imagined you using such filthy language.' I explain they've only ever heard me on telly or radio where I'd get

fired if I didn't rein in my urge to say 'cocksucker'! I was terrified Noel would Gotcha me and I'd swear my tits off, not to mention be impatient and lose my temper. I'm not one for waiting around. So, if Noel had me stuck in a lift or locked in a shop, I'd have been a sweary, furious ball of prickly irritation and put the nation's back up something chronic.

When it happened, I thought I was chairing a debate at Bristol University, 'Are students our future or a waste of space?' on Football League Cup Final day 1996. Luckily, the presence of an audience stopped me turning the air blue when a half-naked posse lurched in, the room plunged into darkness and deafening noises erupted. I behaved with perfect decorum until, after what felt like hours of keeping the beleaguered show on the road, I spotted an audience member, in a woolly hat, football insignia crayoned all over his face, watching a noisy portable television. I was thoroughly fed up. 'Sir,' I said, 'you are causing a disturbance. I'm going to have to ask you to leave.' Sir was Noel, Gotcha hidden up his jumper. I'd been done up like a kipper. He couldn't believe I'd refused to be derailed by his team's disruptive antics, soldiering on and on with the 'debate'.

Punchline

The second Noel handed over the Gotcha – a hideous trophy featuring a defenceless idiot trapped in a giant fist – he strode towards his limo and was driven off at speed. There was no post-Gotcha camaraderie, not even the fake kind, no pizza takeaway, no nod to the fact you'd been made to look a total prat, just a long old drive back to London, hot with embarrassment, dreading the moment your stitch-up would be broadcast to eighteen million guffawing scoffers.

4 I was presented to the then Prince Charles twice. The first time Mr Motivator kept peering over my shoulder, offering ebullient encouragement to steady my wobbly curtsy. 'Yes, baby, balance, baby! You can do it!' The second time, Charles shook his royal head ruefully and said in tones of disappointment: 'Oh, Vanessa, Vanessa, Vanessa.' What had I done to disappoint the monarchy? Get papped demolishing a chocolate eclair? It was eerie. The heir to the throne sounded exactly like my mother.

5 I was a guest at Katie Price's wedding to Peter Andre. It turns out part of being famous is being invited by people you don't know to events you have no business attending. Fame etiquette dictates that when an invitation from strangers arrives on the mat, you don't mark it 'return to sender', you have a butcher's and accept if it tickles your fancy. That's how Martine McCutcheon ended up a bridesmaid at David Gest's wedding to Liza Minnelli. She'd never met bride or groom, but *OK!* magazine thought their nuptial issue would sell better if Martine featured in the wedding party. *OK!* magazine paid £1.75 million for the scoop. A helicopter circled overhead to stop rival publications grabbing aerial shots.

Proprietor Richard Desmond demanded my presence. I knew Peter and Katie slightly and liked them both. That morning every BBC news bulletin bore tidings of the forthcoming knot-tying at Highclere Castle. Katie's £45,000 Isabell Kristensen Disney-princess crinoline was created from pink tulle encrusted with thousands of hand-sewn crystals. It took five sweating staff just to stuff it into the Cinderella horse-drawn coach transporting her to the altar. Kerry Katona was a bridesmaid alongside the late Sarah Harding of Girls Aloud and Liberty X's Michelle Heaton – another *OK!* commandment. Gazza (Paul Gascoigne) somehow managed to cut himself and bleed all over the dance floor. A catwalk carpeted in white and illuminated with neon pink spotlights bisected the room so the happy couple could stride down it en route to the top table. The cocktails were so potent most guests could barely remember leaving at four a.m., after fireworks, a steel band, a biographical light show of the couple beamed on the battlements and performances by Peter Andre's adorable family. I enjoyed myself enormously.

6 The girls and I were flown to Bristol in a helicopter with Bob Geldof, his ravishing French partner Jeanne Marine, his daughters Fifi, Peaches and Pixie, and Jennifer Saunders, with her three daughters, to the premiere of *Muppet Treasure Island*. Landing at an out-of-the-way airfield, we were met by police outriders who escorted our limousine into the city, sirens blaring, lights flashing. Adjusting to normal life among the school-run traffic the morning after took a moment.

7 I bathed in gold-top milk on live TV when the late Keith Chegwin, speciality 'knocking on people's doors and surprising mums in their nighties', did exactly that. He insisted a milkman carry dozens of crates into my bathroom so I could ape Cleopatra and bathe in the Express Dairies' equivalent of asses' milk. I wore a swimming costume, but my gigantic knockers floated to the surface like lifebuoys.

8 I impersonated my own waxwork at Madame Tussauds, standing stock still and bursting suddenly into life, frightening the wits out of innocent tourists. I'm still not sorry.

9 I waltzed with Jeremy Spake from BBC show *Airport* for Children in Need – what they'd done to deserve it, I don't know.

10 I sang 'Rhinestone Cowboy' live with Glen Campbell on *The Big Breakfast* bed. 'You know the words, right, V?' said producer Stuart Murphy. 'Of course! Everyone knows "Rhinestone Cowboy". It's my favourite song,' I said merrily. Glen and I kicked

off promisingly. 'I've been walking . . .' Shit and double shit! I pegged out with a camera right in my face. I could remember something about compromisin' which I knew rhymed with 'horizon' but that bit wasn't for ages. I opened and closed my mouth like Tory MP John Redwood caught shamefully fluffing the words of the national anthem. I apologized to Glen, who graciously forgave me and kindly told a tale about never having visited Galveston, the town in his monumental hit 'Galveston' to create a diversion.

The next week Max Bygraves joined me on the bed. Would I, they wondered, sing along with Singalonga Max to 'Pink Toothbrush'. 'Only if I have every word of the lyrics written in capital letters on cue cards,' I replied. I'd learned my lesson.

11 I was in Shane Meadows's film *Once Upon a Time in the Midlands*. I played Vanessa Feltz, host of the talk show *Face to Feltz*, alongside Robert Carlyle, Ricky Tomlinson, Kathy Burke, Rhys Ifans, Vic Reeves and Bob Mortimer. It begins with Robert Carlyle saying: 'Fuck that Vanessa!'

12 In 2004, I was Marjorie Dawes's guest on *Little Britain*'s 'Fat Fighters'. The plot required a disgusted Marjorie to spit in my face, spraying me liberally with saliva. Matt Lucas and David Walliams are perfectionists. In a hall in the East End, we endlessly rehearsed the dialogue, most of which seemed to be Marjorie, played by Matt, repeating the words 'Vanessa Feltz from the television' in an ambiguously satirical tone. What we did not rehearse was the spitting part. Matt and David were embroiled in technical discussions about the best way to spatter me with genuine spittle without literally spitting in my face. They pondered simply painting on a saliva-esque substance but decided the act of spitting and the propelling towards my face of genuine spittle was essential for maximum comic effect.

Eventually a silicone film, like cling film but three times thicker, was conjured up by the props department and draped across my features. Matt and David were worried. The substance was untried and untested. What if globules of pure catarrh penetrated the layer and landed smack bang on my kisser?

Filming was lush. I rattled about in a state-of-the-art Winnebago. Of course, I couldn't enjoy the amenities. Matt Lucas was about to gob all over my mug. David pacing about the set, feverishly making notes, added to my apprehension. Matt is a grade A spitter. You'll be relieved to know the protective silicone worked a treat. Spittle was sprayed. I survived safe and dry. The sketch is viewed, relished and discussed to this day. Matt will agree he and I have a special bond no one else will ever truly understand.

The ugly side

Fame ruins relationships. In real life, if you have a tiff with your dad or forget to invite an old school chum to your birthday party, they tear you off a strip or stop texting. They might cross you off their Christmas-card list. Whatever they do is confined to you, them and your social circle. It's private. Once you're in the public eye, it's open season. Think about Meghan Markle and the rogues' gallery of grasping relatives peddling gossip for cash. You don't have to marry a prince to be a source of revenue. Anyone who fancies making a buck selling a story about a celebrity can locate a journalist with a cheque book. Imagine you have a distant cousin on Skid Row. He has the choice between sleeping on a park bench or cashing in with some tittle-tattle about you. You can hardly begrudge him trading tales for dosh.

It's different if your dad's a middle-class, golfing, Lexus-driving businessman. When mine hooked up with a divorcée a few days after my mother died and announced an indecently short time later that he was getting married, I didn't feel able to attend. It was much too soon. I couldn't get my head or heart round it. My father didn't seem happy or settled. My mother had barely been laid to rest. In real life, it would have been thrashed out between my father and me. An invitation is precisely that – an invitation. There's no obligation to accept. Instead, my father chose to go to the papers with the 'story'. It seems unbelievable now, but that was the first bad press I'd had. My show had been criticized, but I hadn't. This was the first slur on my character, and it was orchestrated by my own father for money. The 'story' was that I was too much of a prima donna to deign to show up at my joyful father's charming wedding. I must, said the article, be a horrible person to behave so selfishly. My father's marriage, predictably, fell apart but the impact of being publicly dissed by my own dad left indelible scars.

47

Can't we just enjoy it for a moment?

VANESSA CAPTURED A 43 PER cent audience share. That means of all the people tuning in at the time it was shown, 43 per cent watched it. ITV asked Anglia to increase production from three to five shows a week and, gargantuan accolade, scheduled it five mornings a week before Richard and Judy on *This Morning*. We were churning out so many *Vanessa*s (four hundred a year) I had no choice but to bid a fond farewell to *The Big Breakfast*. It was 1997. I was earning enough to pay off the mortgage on the renovated Finchley house. We'd wrestled with crushing mortgage payments since our wedding day. Now, two decades earlier than we could ever have hoped, we were free. My relief was intense. We'd moved four times in ten years of marriage, trying to transform rancid properties on a budget of nil through sheer determination, always prevailing upon the generosity of Mr Barclay, a wing and a prayer. For the first time ever, we weren't weighed down by sleep-depriving debt. Hooray. Let the good times roll.

I envisaged a steady flow of piña coladas, sipped from a sunbed beneath a coconut tree. Not really, but I did imagine a few parties, a bit of guilt-free fun and maybe putting a down payment on a flat each for the girls that they could either live in or rent out when they grew up. I'd been grafting away since the age of seven, slicing through my hand with a Stanley knife in my dad's underwear warehouse, being stitched up at Edgware General Hospital and straight back into the warehouse packing brushed-nylon pyjamas all afternoon. I had no intention of working any less, but I did have every hope of enjoying the fruits – nay, a resplendent physalis-adorned fruit platter – of my labours. The Good Doctor was poised to become a consultant in the NHS. He'd be able to set himself up in private practice, carve a niche in knee replacements and earn a few quid. I was gearing up for at least a couple of afternoons of bacchanalian pleasure, attired in a YSL caftan, reclining on a lavishly cushioned chaise longue.

Not a chance

The Good Doctor had other plans. The girls were splashing away in the pool. (I spent some of the money from series one of *Vanessa* on having one put in, thinking if it didn't get recommissioned we'd all enjoy it for years to come.) He was luxuriating in the raised corner bath in the obscenely expensive en-suite he'd personally designed, featuring special marmalade marble hand-hewn from the vaults at Carrara, crystal chandeliers, hand-woven Brintons carpet with a decorative border and walls delicately hand-painted with *trompe l'oeil* faux marmalade marble. I swear I'm not exaggerating a single detail. You might be wondering how and why I countenanced such overarching extravagance. Let me try to explain.

Whatever you do, Vanessa, don't break the Good Doctor's balls

I was a target for an avalanche of unsolicited advice. 'How does your husband feel about your fame?'; 'The key thing is to make him feel he's wearing the trousers'; 'Don't break his balls'; 'Don't call the shots'; 'You may be out-earning him, but whatever you do, don't show it.' Everyone agreed. Don't throw your (considerable) weight around. Let him be the Boss. When I say everyone, I include myself. Of course I agreed. The Yiddish chorus was correct. I knew my crucial job – significantly more important than my day job – was to play down my achievements, especially my pay cheques. The last thing I wanted to do was surpass my husband the doctor.

The shtick in our marriage – and I was happy to foster what I now know is fiction – was that his was the 'real' career, the serious, meaningful, important one. He was saving lives and making sick people better. My job was inferior, superficial, vulgar and wouldn't last. He told me regularly, with conviction, my career would survive a maximum of seven years. I believed him, though in retrospect I don't know why. He was a surgeon, not a soothsayer. How could he possibly know how long I'd be employed?

I'd say: 'I'm thinking of buying Allegra a pair of shoes' or 'Do you think we could stretch to a trip to Barcelona for the Easter weekend?' I'd never say: 'I'm doing A or buying B. It's my money and I don't have to ask anyone's permission.' I was trained to create *shalom bayit*, peace in the home. If that meant channelling my 'Jewish geisha' and making sure the Good Doctor was in charge, it was a small price I willingly paid.

We move house for the fifth time

The Good Doctor said we should move house. 'Move? What? Why on earth would we do that? This house is fantastic. We've done up the whole place. Butter-yellow kitchen. Ill-placed walls bashed down. Both girls' rooms pink and pretty. Two pristine bathrooms. Amtico floors. Swagged and tailed curtains lined and interlined. An office with his and hers desks. The pool, for Heaven's sake. We've only been here two years. You've barely cut your toenails in the en-suite. This is the first time we've been debt-free in our whole marriage. Why would you want to move now?'

His reply was disdainful. I didn't recognize the tone. 'Don't you understand? That's how the property market works. You gear up. You extend. We can afford a bigger, more prestigious home. It would be a criminal waste not to take advantage of the opportunity. You don't get it. Only fools sit tight on a mountain of cash. It would be moronic not to make our money work for us.'

I tried gentle dissuasion. I didn't go in hot and heavy, flatly refusing, because I didn't want to be a bossy headmistress cramping the Good Doctor's style. Tentatively I did my best to resist. 'We're so happy here. Can't we stay another couple of years?' No dice. He was busy, looking for a *shprauncy* (showy) new house – bigger, better, swankier. Seriously, the man was looking at embassies. We nearly made an offer on the Cameroonian Embassy in Stormont Road, N6. We even paid a king's ransom for a survey on the rambling mausoleum. Luckily, in the nick of time, we were tipped off that the chap 'selling' it didn't own it and was hoping to scarper with our cash. We settled on a 10,000-square-foot, crumbling mansion in Winnington Road, N2, parallel to Bishops 'Billionaire's Row' Avenue.

The house lacked fundamentals: front door, half a roof, a boiler, floors, eleven baths. Yes, there would be eleven bathrooms when someone bought and installed them. The developer went bust part-way through the renovations. The builders downed tools. Wind whistled through what was nothing but a vast building site. A fox lived in the loft. We agreed we'd purchase this money pit for a gut-churningly astronomical sum, which would include the absent builders completing all the unfinished work. We had some cash from selling the previous house, but the mortgage would be £1.6 million. The thought made me nauseous.

On the day of exchange, the estate agent called. 'A Russian with a suitcase of money' had appeared on his doorstep, raring to gazump us. Real and mythical Russians touting Louis Vuitton cases bristling with crisp oncers were all the rage in 1997. London was the world's leading money-laundering city. I don't know if our Russian was fictional or flesh and blood, but the Good Doctor believed the tale and was determined not to lose our half-built, mega-pricy mansion. He decided we'd purchase the house for the agreed price BUT AS IT WAS – unfinished and uninhabitable! OH, MY GOD! We were somehow going to have to find the money to make the place watertight and warm, with a roof, boiler, brace of bathrooms, and pay the swingeing mortgage on top. Did I mention he was on the ski slopes, slaloming down an Alp while barking orders at me? 'Exchange! Do whatever you have to do! Don't fuck it up and lose the house!' After picking up a notarized document from a minion positioned beside a holly bush on the Brent Cross roundabout and several zigzags to and from the West End, the deed was done.

I took on another gig opening Thorntons chocolate shops all over Britain every Saturday. I was famous but penniless. Every pound I earned went straight into the cement mixers ceaselessly churning away at the new house.

48

Life in a rock-star mansion

I DIDN'T KNOW IT TILL I did it, but gearing up to be flossing, flossing and glamorous is a breeze. I took to it like a natural. Ten minutes as the chatelaine of the rock-star mansion and I was sashaying about as if I'd been swanning around in ten thousand square feet of splendour all my life.

Normal things looked stupidly small in such sumptuous surroundings. The marital bed, a standard five-footer, had to be replaced with a seven-foot megalith complete with built-in contraption from which a gigantic telly was magically raised, then lowered into a bespoke walnut cabinet. The master bedroom, complete with not one but two en-suite bathrooms, dressing room and wooden floor inlaid with an exquisite mosaic, was tennis-court-sized. Two sofas, two armchairs, a table, two hand-carved bookcases and a magnificent dressing table disappeared into the voluminous space.

The indoor pool, in its own complex, with showers and changing rooms, was located off the kitchen and contained our monogram in swirly mosaic, finished off with understated cavalcades of dolphins.

The Good Doctor had a majestic office, decked out in musical-manuscript wallpaper and fabric. I had an office, decorated in primary colours like the *Big Breakfast* set. The girls each had a bedroom and bathroom, Allegra's bouncing with bows and ribbons, Saskia's themed under the sea with 3D treasure and a lost diver floating down the wall. They shared a vast playroom decorated with murals paying homage to Barbie, Polly Pocket and Sylvanian Families.

We had a ballroom, for Christ's sake – an actual freakin' Cinderella-esque ballroom. I had star cloth fitted on the dining-room ceiling, Orion's belt reflected in the black marble floor. The whole house was stratospherically, ostentatiously, excessively extra – the perfect venue for the party of the decade.

I bid you a cordial welcome to . . .

Allegra's batmitzvah

How do I describe the joy of throwing a shindig to celebrate your beloved daughter's coming of age? There's only one word to do it justice. It's Yiddish and it's *nachus. Nachus* means the heart-swelling sensation of pride and gratitude generated by your children (and grandchildren). When the *nachus* reaches saturation level another Yiddish word kicks in. It's a verb, to *kvell. Kvell*ing is what you're doing when your adoration of your children is so overwhelming your mind, body and soul are consumed by love. You are not merely living, watching, enjoying. You are *kvell*ing. Allegra's batmitzvah was a *kvell*-fest. Did I happen to mention both my daughters are brilliant and beautiful?

OK, batmitzvahs are a sop. They're a weak and feeble substitute for the real thing – barmitzvahs, where boys become men, strictly testosterone only. Batmitzvahs were invented in the seventies to keep bra-burning feminists off rabbis' backs. Boys mark their ascension to manhood by reading from the *Sefer Torah* (holy scroll) in front of the sacred ark at the heart of the Sabbath service. Girls are relegated to the synagogue hall, which always reeks of chopped herring, kept a million miles from the scrolls in case they menstruate in their vicinity and made to deliver a *d'var torah*, which is Hebrew for a trumped-up speech at the whim of their *cheder* teacher at some ungodly time in the afternoon.

No one takes the girls anything like as seriously as the boys except the parents of daughters only. Not having sons to celebrate, we were wholly immersed in marking Allegra's passage to womanhood. I didn't fancy being siphoned off to rubbishy old Sunday teatime. Hours of negotiation allowed us to commandeer the synagogue hall for a small spin-off service on a Sabbath morning, and although Allegra didn't get within a hundred yards of the ark, she delivered her *d'var Torah* exquisitely. At least the rabbi bobbed in and dispensed a blessing to kosher the ceremony up a bit.

Obviously, Allegra and all her discerning friends realized the 'bat' was in every way subordinate and inferior to the 'bar'. How did we defend the indefensible? We did the spiritual thing, of course, and focused heavily on the party. *Hello!* magazine devoted eight pages to the festivities. Laser beams guided our guests and pierced the sky all the way to the Almighty on his perch in Heaven. The *Jewish Chronicle* plastered posters all over London: 'Vanessa Feltz Is Having An Affair'. One hundred and ten revellers were seated in the ballroom, plus Banana Split – the disco and two dancers – and a rock-and-roll band. Fifty twelve-year-olds sat at trestle tables on our ice-rink-sized upstairs landing. The caterers prepped in the kitchen, garage and pool-house. One over-excited waiter slipped and ended up underwater, tray of vol-au-vents still in hand. *Shomers* – pious men paid to check the kosher caterers are scrupulously kosher – combed the exotic fruit buffet for non-kosher ants and greenfly.

Allegra, angelic in pale pink beaded chiffon, was the belle of the ball. Mid-speech, she subsided into sobs, overcome. Saskia, nine, a dream in white satin, stepped up, put her arm round her sister and took over.

I was sublimely, supremely happy, proud of my daughters, in love with my husband, relishing my unexpected career. I missed my mother. She would have added another layer of adoration and entirely appropriate Allegra-worship (she was a critical mother but a besotted grandma). I appreciated every particle of my good fortune. That night was one of the happiest of my entire life.

49

Hello, Vanessa, this is the BBC

I DIDN'T DO A MICHAEL Barrymore and make my chauffeur drive me round the West End after dark while I stood upright in my Rolls-Royce convertible, waving at strangers, but by the summer of 1996 I did have a chauffeur. The late lovely Stephen Blumenow was the first kosher chef on the *QE2* and when he wasn't driving me up and down to Norwich in my newly acquired second-hand dark purple XJ6 with cream leather seats he made ravishing chocolate mousse.

Everything ticked along splendidly until, sitting at home minding my own business on the morning of 26 June 1998, I answered the phone to the BBC. Auntie had been watching me. Auntie wondered if I was appreciated and understood at ITV. Did the channel value my intellect and versatility? Was I being given the chance to play to my strengths?

Remember what I told you before about 'talent-handling'? Well, ITV weren't bothering. There'd been no handling at all. I'd been making *Vanessa* in Norwich for five years, and not once had a representative from ITV even visited to say hello. I'd never met any of them. Channel 4 gave me *The Big Breakfast*, BBC1 *Value for Money*, Channel 5 *Vanessa's Day With . . . ?*. ITV, for whom I was making five humdingers a week, ignored me. I tried to make an appointment with ITV controller David Liddiment. He granted me an audience, but when I walked into his office, he said he had to dash to pick up his parents at the station and could give me three minutes. Was there anything in particular I wanted?

Naturally, the BBC diplomat bombarding me with blandishments had my undivided attention. I had been dying for someone powerful to observe: 'Vanessa, you are fabulous on the show, but we're under-using you. Weren't you at Cambridge with Stephen Fry? Could you present an incisive documentary series on logical positivism and a nightly quiz on Wittgenstein?' This fellow from the BBC was saying everything I'd most wanted to hear. Would I agree to a clandestine meeting on the terrace of the Halcyon Hotel? What do you think?

Camouflaged behind a bush were BBC1 controller Peter Salmon and head of daytime Jane Lush. They said they adored everything about me. They were desperate for me to present a daily talk show for them, plus a Saturday-night prime-time show and features. 'What Carol Vorderman's done for maths, we want you to do for literature at the BBC. There'll be no more driving up and down to Norwich. We'll build you a beautiful set at TV Centre just up the road from your house. ITV don't really get you. We do. You are unique – a sharp intellectual with the common touch. You are squandered on ITV. Come to us and we will help you soar.'

Obviously, I knew the compliments were a strudel of schmooze. I realized they were saying whatever they thought I most wanted to hear. It didn't stop me wanting to hear it, though, or lapping up every word. Money was mentioned. Monopoly money. I was already earning a sizeable stack at ITV, but this was big bucks. It had to be because they were proposing what was called a 'golden handcuffs' contract. Once signed, you can't work for any other company.

I relayed the phenomenal tidings to my solicitor, Graham Shear. 'Enjoy this!' he said calmly. 'Savour the moment. It'll come to nothing. It takes for ever for the BBC to ratify anything. You're back at ITV at the beginning of September. The BBC will never make this happen in July and August. Be pleased they paid you this compliment, then forget about it.' I told the Good Doctor about the wonderful offer and what a shame it was that nothing would come of it.

The Beeb moves like greased lightning

Except it wasn't and they did. The BBC rushed the proposal through at an emergency rate and, within three days, the deal was ratified and the documents waiting on Graham's desk. There was just one problem. I was under contract to ITV and the BBC stipulated ITV must not know they'd issued me what's called a 'What If?' contract. It was waiting in a drawer but must be kept top secret so Auntie couldn't be accused of poaching me. Graham had six weeks to winkle me out of ITV without them catching on. He began demanding the impossible, digging his heels in, insisting, for example, I'd only make the show in London. ITV couldn't believe their usually compliant presenter was suddenly asserting herself.

If you're wondering why I wanted to jump ship to the BBC when I was so happy at Anglia, it was partly because I grew up in what used to be known as 'a BBC household'. We were *Blue Peter* not *Magpie*. We venerated the BBC. My parents considered ITV 'common'. I also wanted to work near my home in London, for an institution respected all over the world, making more challenging programmes. I loved Anglia and *Vanessa*, but I'd made over a thousand *Vanessa*s in five years. I was only thirty-six and felt ready to spread my wings. Plus, there was a ton of money in the

BBC kitty. It would make mincemeat of the fearsome mortgage. In fact, this was 1998 and if Graham could get me to the finish line, I'd be able to pay off the entire mortgage by June 2000.

Graham Shear is, as his numerous illustrious clients will testify, possessed of magical powers. In the nick of time, after weeks of jousting, he extricated me from my ITV contract. One minute later, I signed the 'What If?' document binding me to the British Broadcasting Corporation body and soul. I dashed to tell the Good Doctor. Was it my imagination or did he seem less elated than I'd expected? Maybe my frivolous foray into television didn't seem quite as preposterous with a rainbow's-end pot of gold attached and a BBC seal of approval. I was doing a foot-shuffling 'Yay! This is so exciting!' dance. He looked disapproving. I should turn down the volume, tone down the fuss. I didn't get it. I wasn't leading a ticker-tape parade down Regent Street, just celebrating fantastic news with my husband in our hotel bedroom. Once again, I'd been too much, too full of myself. The feeling was familiar. It was exactly like being disapproved of by my two most fervent critics, Valerie and Norman Feltz.

Was I puffed up with pride? I don't think so. I think I was bloody relieved, flattered and dying to pay off the crucifying mortgage. Maybe I was proud, guilty of overweening hubris. I must have been because Humpty Dumpty and I have a colossal amount in common.

50

The great fall

Everything went horribly wrong.

1. ITV didn't do the usual courteous thing and release a statement saying: 'Vanessa Feltz will no longer be presenting ITV's *Vanessa*. We wish her every success in her future endeavours.' Instead, they put out a venomous press release saying they had terminated my contract because I'd asked for a huge pay increase. I hadn't. Graham Shear's demands pointedly did not refer to money. The ITV statement was inaccurate, but I couldn't counter it by revealing I'd left ITV to go to the BBC: the Beeb still wanted my contract kept secret so they could avoid any accusation of poaching ITV talent. I couldn't say a word.

2. The story was front page and it stuck. It made sense, especially if you were just a little bit racist. I was fat. I was Jewish. The idea that I was a big fat Jew who was greedy for money fitted every ancient anti-Semitic trope. No one suggested I'd drunk the blood of a Christian baby, but I felt it was just a matter of time till they did. The coverage was vituperative. Just who did this woman think she was? I could have answered that. 'Nothing special. It wasn't my idea. The BBC laid siege to me, and I went for it.' I wasn't allowed to say that, but it was clear no one would have believed me if I had.

3 The BBC took far too long to announce my signing and launch their version of *Vanessa*, to be called *The Vanessa Show*. It was September and their leisurely plan was to air in January. Meanwhile Malcolm Allsop instantly put the feisty Trisha Goddard on my old set with my audience to present what, days before, had been my show. Trisha, now a good friend, hit the ground running. She's straight-talking and empathetic. If you liked the format, you were happy watching Trisha. Within days it was 'Vanessa who?'

4 *Kilroy*, a stale debate show, hosted by the perma-tanned ex-MP Robert Kilroy-Silk, had been dragging down the BBC's morning schedule. ITV's *Vanessa* was pulverizing it in the ratings. It was obvious the BBC wanted to sign me so they could get rid of *Kilroy*. It was dead wood. Everyone who cared knew the first thing the BBC would do on securing my services was ditch *Kilroy*. Only they didn't. Lord knows what BBC bosses were thinking. Unfathomably, they hung on to the show. *Kilroy*'s few fans were getting their fill on BBC1. Ex-*Vanessa* fans were being comfortably served by Trisha on ITV. There was no vacuum with my name on it. Viewers had an elegant sufficiency of talk shows without me.

5 It took the BBC four months to get their anaemic *The Vanessa Show* on the telly. At the secret meeting on that hotel terrace, Salmon and Lush assured me they gloried in the red-blooded vulgarity of my ITV show. They cited 'Get Out My Bed! I Want a Man From the Med!' as a favourite. They couldn't wait, they said, to expose the sheltered BBC audience to my rumbustious sauciness. I believed them. To me the speedy contract and thwacking cheque signalled sincerity. They signed up most of my Norwich team.

6 Fish out of water at TV Centre, my Anglia colleagues looked lost and worried. The bosses were pasteurizing programmes. Ribaldry

was being replaced with respectability. I'll give you an example. The BBC insisted my guests' tattoos must be hidden. They didn't want to feature folk who were too obviously working class. Marks & Spencer shirts were kept by Wardrobe especially to mask my guests' 'unsightly tattoos and piercings'.

7 I was terribly worried. The funny, sexy shows we'd been making successfully were being sanitized until every ounce of vitality was squeezed out of them. The topics were dull and worthy. The guests were serious and middle class, not at all our usual *Canterbury Tales* crew. We knew how to make our shows soar, but our wings were clipped. We realized the pilots we were making were much too genteel for the genre. We could have rescued them, but we weren't allowed to. My editors, producers and researchers had uprooted their lives and moved across the country to stuff a firework up the backside of BBC talk shows. Instead, we were forced into making a flaccid flop of a show and the public – underwhelmed focus groups apart – hadn't even seen the wretched thing yet.

8 Controller Peter Salmon wouldn't take my calls. I didn't know what to do to stop my show dying an agonizing death before it had even begun. In despair, I took to writing Salmon lengthy letters and handing them to his secretary. 'Please, please, please listen. We know what we're talking about. This is what the show needs. Please allow us to do it.' I'd list the SOS procedures necessary to triage the programme and bring it back from the brink of death. There was no response.

9 By the time the BBC *Vanessa Show* went on air, the press pack were pawing the ground, sniffing for the whiff of failure. Vanessa Feltz was being paid a king's ransom of licence-fee payers' money to inflict a vulgar ITV import on innocent BBC viewers. She'd better bloody well deliver. What they didn't know, of course, was

that we were about to serve up fare so anodyne it wouldn't offend Hyacinth Bucket.

10. The show opened to underwhelming ratings. The tabloids claimed I'd boasted, 'I'll trash Trisha!' I hadn't. The story was invented. First, I've never used 'trash' as a verb in my life. Second, I hated my pallid apology for a show. I'd much rather have watched Trisha's. Anyone with an ounce of sense would have felt the same. It didn't stop the hacks lambasting me for being tragically trounced (a verb I do use) by Trisha. She was wiping the floor with me.

11. The Good Doctor didn't show any interest. I was being pilloried in the press. The show was a snooze-a-thon. Management wouldn't countenance a resuscitation plan. My team were mutinous. Thank God social media didn't exist to make the embarrassment even more toe-curling.

12. He decided to throw a casino-themed party for his fortieth birthday. Scantily clad croupiers and wasp-waisted waitresses wiggled about ladling champagne down friends' throats. I was humiliated. I didn't want to show my face let alone leap about the dance floor.

13. After all the fanfare and fandango, the move to the BBC was an ignominious failure, a big blobby blot on my career copybook. It was a formidable fuck-up. It was all my fault for falling for the BBC's blandishments and jumping ship and my closest confidant, the Good Doctor, didn't seem to give a damn.

14. Going live on air delivering a show no one likes, not even the host, is soul-destroying. I couldn't sleep. I had anxiety nightmares – the ones where the water's closing over your head, you're gasping for breath, and when you snap awake, you pant, holding on to the mattress, listening to your heart thumping.

51

That was nothing

Everything was about to get a whole lot worse.

I was working in my office at the front of the house when Anne Morrison, a BBC boss, rang. 'Vanessa, I have bad news for you. Tomorrow the *Mirror*'s front page will read, "VANESSA IS A FAKE!" There have been fake guests on your show. This is incredibly serious. Await further instructions. Do not, repeat not, discuss this with anyone.'

'VANESSA IS A FAKE!' I stood up in shock. In my front garden, squeezed among the rhododendron bushes, were paparazzi, reporters and film crews. The girls were about to be dropped home. If I stepped out to meet them, I'd be caught on camera. If I stayed inside, they'd be scared witless. I rang my husband. Something awful had happened. Could he come home to be a reassuring presence, please? He said he'd try to pop back at some point but was going to be out that night 'playing bridge with the boys'.

'Is there any chance you might rearrange it? I'm not sure what's going on and it feels ominous. Could you please just be here with me and the girls?'

Reluctantly, he agreed to try to re-route the bridge game to our house. He sounded aggrieved. The undertone was: 'You see? Piss around attention-seeking on television and you get what you deserve. What did you think would happen?'

The answer – which I didn't give of course – was 'Not this!'

If there had been fake guests booked on the bland BBC show, it was the responsibility of BBC Productions. Presenters don't book guests. We have to trust the producers who tell us about people we interview. If they say, 'The first guest is Susie. She's a single mum from Rochdale who deals drugs to pay her daughter's private-school fees,' the presenter must believe Susie is who they say she is. If she turns out to be a motor-racing bird-watcher from Scunthorpe, it's news to us.

I was told there had been a couple of shows featuring guests who weren't who they said they were. Two sisters weren't sisters but strippers who'd been paid to pretend. Live on the show, I had questioned whether the two women were related. They looked nothing like each other and had completely different accents. 'You don't seem like sisters,' I said.

'She's like our dad. I'm like our mum,' they replied, along with some waffle about being brought up in different countries.

Lead story on the BBC Nine O'Clock News

'Fake guests on *The Vanessa Show*' was the lead BBC1 *Nine O'Clock News* story that night. A reporter stationed outside my dressing room solemnly intoned, 'This is the dressing room in which Vanessa Feltz of *The Vanessa Show* prepares to greet her guests, several of whom turned out to be fake.' Just six and a half months after the BBC ardently pursued me, my show was branded the factory of fakery. It bore my name. Both were tainted. Pundits talked of scandal, traducing Reithian values and the cancelling of the licence fee. Something was rotten in the state of British Broadcasting – Vanessa Feltz.

Watching alone in the bedroom, I wished my husband might take a brief break from bridge and hold my hand. He didn't. When, hours later, he came to bed, he exuded irritation. I'd made my bed. I'd better lie in it. Fame is a poisoned chalice. Surely I must have known nothing good would come of it.

The Vanessa Show became a battering-ram to attack the BBC. Newspapers ran the story day after day after day. I was made to visit the *Mirror*'s offices and abase myself in front of the then editor, my now friend and erstwhile TalkTV colleague, Piers Morgan. BBC boss Matthew Bannister asked Piers to produce a tape he claimed the *Mirror* had in its possession of one of my team knowingly booking fake guests. Piers refused. Members of our studio audience were required to produce birth certificates to show they were bona fide, not, Heaven forbid, fakes. Three staff were dismissed. They were my friends, not merely co-workers, and I didn't believe then, or now, that they'd booked fake guests deliberately. I think they were scapegoated to save the BBC's tarnished reputation. Three stern sergeant majors from the BBC's old guard were put in charge to ensure there were no more irregularities. You can imagine what a fillip they were to the already turgid editorial content. The show was heading for the knacker's yard. In July, Peter Salmon, the eager swain who'd wooed me the year before, announced: '*The Vanessa Show* was a boil and we've lanced it.' RIP the BBC *Vanessa Show*.

52

That was nothing

THE GOOD DOCTOR'S BROTHER DIED in his late forties. It was a family tragedy. My husband was absent more. I put it down to his grief. He seemed distracted. Odd things started happening. Without asking me or telling me, he revved home in a spanking-new car. He'd always said new cars were an absurd waste of money. 'By the time you drive them out of the showroom, you've already lost a fortune.' He'd always said: 'Our family only buys cars second hand.' This one was a sporty two-seater. We had two children aged ten and thirteen. What were they supposed to do? Get the bus? He'd ditched the personalized number plate he loved DoCT3R. He adored the plate. Where was it?

'What happened to your number plate?'

He replied scathingly: 'It's just not who I am any more.' He said it as though I'd have realized this, if I'd had a brain. I felt stupid for not grasping he'd changed.

Knowing we had twenty-seven people for dinner to celebrate the Jewish New Year, he disappeared without trace and didn't show up till our guests were halfway through the soup. There was no explanation. On the Day of Atonement, when families go to synagogue together, he said he didn't feel like coming. The Christian equivalent would be your husband saying he doesn't feel like joining in Christmas dinner. He stopped watching television in bed with me and started reading a book on personal development. I was floundering without my show, not sure what I was supposed to do every day. He wasn't interested. I put his strangeness down to the loss of his brother. I felt for him and tried to let him know I loved him and would support him in every way I possibly could.

You might have been wondering what my husband thought of my weight. He was slim and fit. How did he feel about being married to 'the woman who ate her audience'? Obviously, I can't tell you what he thought. All I can do is tell you what I thought he thought. I thought he thought I was lovely. I thought he thought I was funny, kind-hearted, hard-working, a good mother and a loving wife. I thought I was lucky to love a man who loved me back. I thought he thought we were together for ever, through thin (him) and thick (of waistline, me). I don't suppose I thought he preferred me five sizes bigger than I was when we met. If you'd asked me then, 'Would your husband like you slimmer?' I'd probably have said, 'Probably?' He didn't seem concerned about my weight, though. He never mentioned it, and I thought I'd been supremely fortunate to find a man, *the* man, who – forgive the schmaltzy cliché – loved me unconditionally.

I was happily married. I loved his quick-thinking problem-solving dynamism. I loved it when strangers came up to him in the street and thanked him for repairing their broken hips and dislocated shoulders. I loved his piercing blue eyes and the dimple in his chin. I loved him very much and thought a whole life by his side would never be long enough.

53

I do not rule out the possibility of a divorce

One September Sunday morning I was peeling potatoes. We were expecting my mother-in-law for lunch. The girls were at Hebrew classes. It was the most ordinary of Sunday mornings. Our pals, the Lewises, had just had a baby, sixteen years after their last. We were going there for tea. I wondered aloud whether we'd cuddle their bouncing bundle, have a sniff of her neck and decide to try for another baby ourselves. I was only thirty-seven. We certainly had plenty of bathrooms! Allegra and Saskia were thirteen and ten. They'd love a baby to play with. 'Who knows?' I said. 'We might feel broody.'

The Good Doctor replied, in a staccato Dalek voice I'd never heard before: 'I – do – not – rule – out – the – possibility – of – a – divorce.'

What? Hang on! What was that? Why was he speaking in that weird way? What had he just said? It couldn't possibly be what I thought he'd said. Could it? He couldn't have casually referred to 'the possibility of a divorce' without once mentioning there was anything wrong with our marriage. I must have misheard. I must have got it wrong. He didn't say it again.

The day rolled on like all our normal Sundays. We picked up the girls, ate lunch with his mother, visited our friends for tea. It wasn't till the girls were bathed and in bed I finally had the chance to ask: 'You know what you said this morning? What did you mean?'

Again, the Dalek voice: 'I – do – not – rule – out – the – possibility – of – a – divorce.'

'Darling, what do you mean? Are you ill? Are you OK? Can I help you? What is happening? I love you. You love me. We love the children. We are so happy. What are you saying?' I was crying, shaking, trying to hold his hand.

His reply was emotionless, glacial: 'I – want – a – divorce.'

'But why? Why? We are so happy. What do you mean? What has changed? You can't mean this. Not really. Not at all. Why are you saying this, my darling love?'

His answer was brief. 'You are just so fat, so fat. It's hideous. You are hideous. I keep waiting for you to get diabetes.' He went to bed in a room on another floor.

I didn't understand. I couldn't breathe properly. I could hardly see. I wasn't me. If I didn't have a husband, my husband, the man I loved, who loved me, I was nothing. This couldn't be happening. He couldn't want to leave – not without arguments, growing apart, years of counselling, acrimony, recriminations, misery and finally, reluctantly, extricating each other from the torture of a rotting marriage. We hadn't done any of that. We'd been rooting for each other, cuddling, pasting and hanging wallpaper in five different houses, grafting, saving, celebrating. Surely nobody living a jolly, companionable married life with two delightful little daughters announces their marriage is over and clears off without further explanation?

We've seen *Kramer vs. Kramer* and *The War of the Roses*. We know how divorces happen. There is corrosive long-term wretchedness. You must try to rescue your crumbling relationship. You don't just walk out. When we danced at weddings he'd croon along with Billy Joel to 'Just The Way You Are'. I believed him. I thought he saw past the fat. I didn't think he minded. He never said he did. I didn't realize he did. If I'd known, I wouldn't have evened off those cakes. I would have tried harder.

I am prepared to give you a trial period

A few days later my husband said: 'I am prepared to give you a trial period. Twelve weeks.' I didn't know what I was 'on trial' for. I asked but he wouldn't say. He barely spoke at all. My clue, of course, was 'You are so fat, so fat.' I had to win the trial. I had to save my marriage, keep my daughters' father in our family home to raise and nurture them. This wasn't a challenge I could fail. There was no margin for error. I thought: Right. I must get as thin as I can as fast as I can so he can see that underneath the fat I have bones and features. He will remember I am pretty, and he will love me and not leave me and the children.

If I didn't want to lose my husband, I had to lose weight. I ate fewer than three hundred calories a day – a few apples, Diet Coke, a hard-boiled egg. I wasn't hungry. I was terrified. I'd never been near a gym since school. I booked a personal trainer. A Jamaican-Mancunian who brought so much positive energy to the project, I called him the Angel of the North. I trained every day, seven days a week. The weight dropped off. I lost a stone the first week, nine pounds the second. I was taking up less space. I filled the house with friends and family. I had soup bubbling on the hob, laughter bubbling round the table, hope bubbling in my heart. The pounds were falling off me and look how happy we were. Our gorgeous girls were giggling. Our mates were cracking jokes and drinking wine. Our fourteen years together had built this warm, welcoming, harmonious home. He wouldn't leave all this. He couldn't leave all this. No one in their right mind would pack a suitcase, walk out the door and turn their back on all this love.

Six weeks into the trial we went to my cousin Simon's wedding. I was wearing a beaded halterneck, black and white Cynthia Raffa-Rickman evening gown. It had arrived in error. It was an American twelve, a UK sixteen. I hadn't got round to returning it. Now, I easily slipped into it. I had my hair swept into a 1999 Denise van Outen-inspired choppy updo. Wedding guests were amazed. They said so. They were expecting me to be fat. 'My God! You've lost so much weight! You've disappeared.' Disappearing was the goal. I was obviously succeeding. For the first time since the trial began, I started to believe my husband might not leave. He seemed more amenable. He danced with me and held my hand for 'Auld Lang Syne'. He was chattier than he'd been for ages in the car home. He got into bed. I sat at the dressing table taking pins out of my hair. This was my darling Good Doctor. This was our married life together. The constriction in my chest lifted slightly.

Suddenly he sprang out of bed, started dressing, grabbed a suitcase. I begged him to stay, went down on my knees, held on to his ankles, was dragged along the acres of gleaming marble. I thought of throwing myself in front of his car as he accelerated away. He shouted over his shoulder he'd be back in the evening to tell the girls. Next day, on autopilot, I taped a show with Dale Winton. Nothing seemed real. At six p.m. the Good Doctor strode up the stairs. Allegra and Saskia were watching television on our bed. The phone rang. He answered it and had an upbeat conversation with one of my girlfriends. Then he switched off the set and announced to his children in the eerie Dalek voice: 'Your mother and I no longer love each other. We will no longer be living together. I am moving out. I shall be buying a flat.'

'Wait! Wait!' I wanted to howl. 'That's not what departing dads say. We've all seen enough films to know the acceptable script by heart. Where was the part about "I'm leaving your mother, not leaving you"? Where was the bit about "I love you with all my heart. I'll be back tomorrow to take Saskia horse-riding, on Wednesday to take you both out for dinner and on Friday to take you to Grandma for Friday night dinner"? Where was "This has nothing to do with you two. You are my beautiful, kind, beloved daughters. This is a grown-ups' thing and doesn't mean I won't be part of your lives all the time for ever"?' This dispassionate bulletin was not, in my view, the way a father should break the shattering news of a divorce to his daughters. He was already heading for the stairs. My daughters were screaming – not crying, screaming. Screaming. Screaming. Screaming. It was as if they had been stabbed. In a way they had. I will never forget. I will never forgive.

I was a woefully inadequate parent because I shared every scintilla of the girls' shock. I hadn't been able to warn them or insulate them against a divorce because I'd been blindsided. People unhelpfully said: 'It's the children I feel sorry for. You never know what goes on in a marriage behind closed doors.' Well, I hadn't known what was going on in my marriage. I hadn't had an inkling. I hadn't been turning a blind eye or pretending I didn't see signs of trouble. I hadn't seen anything.

Is it vulgar to point out that, ever since signing my BBC contract, I'd been paying off the material sum of our mortgage at a sizeable thwack every month? OK, I know it's vulgar, but I'm telling you for a reason. Don't you think maybe if your wife is coughing up that kind of money to decrease your mutual debt you might be extra careful to make sure she thinks you're happily married so she carries on doing it? If I'd had the shadow of a doubt about our future together, I'd have stopped the payments immediately and secured the girls' financial future by buying them each a flat in their name.

I never suspected a thing because the Good Doctor was meticulous in making sure I saw nothing suspicious, until – and this is only my opinion but it is shared by my lawyer and financial advisers – he saw my BBC show axed, decided the goose that laid the golden egg wouldn't be laying any more, and there was no point in hanging around any longer. The show was axed in July. He legged it at the beginning of October.

54

I've never felt so cold

How do I describe the shock? You know by now I'm not short of words. It's just hard to find the right ones to sum up the desolation. The person I wanted to turn to for comfort was my husband. He'd disappeared. I couldn't work out why. I couldn't explain it to the girls, give them a plausible reason, reassure them they were going to see him again or tell them when because I didn't know, and he wasn't communicating with me. Someone said he was 'staying at his mother's'. He was anything but a mummy's boy. He could easily afford to rent a flat or book a hotel room. What was he doing in his mother's back bedroom? The girls' sadness was etched on their faces. I found them a therapist each, hoping they'd be able to discuss their feelings freely without worrying about protecting me or each other. I made appointments with their head teachers.

Before the Good Doctor left, he muttered something about 'making a statement to the press'. Who did he think he was? Brad Pitt? I'd never given the press the lowdown about my private life. I didn't want to do so now. The ladies and gentlemen of the tabloids swarmed. They followed me to a family funeral. They tracked down me and the girls at the supermarket. They were permanently camped in front of our drive. They pointed lenses through our windows. On the coach to school Allegra heard Chris Tarrant, on the radio, shooting the breeze about her father leaving. Jonathan Ross was a fellow parent at Saskia's school, King Alfred's. Spotting me, he dashed over. Putting a comforting arm around me, he took me for coffee in Golders Green. He was exceptionally kind. Despite the abrasive persona he projects on TV, to me he was a Mandela–Gandhi mash-up. I'll love him till my dying day.

Speaking of dying, I thought I might. I know the difference between figurative and literal, and I felt as if my heart was literally broken. It was as if I had to wait, without oxygen, for the blood to pump from one chamber to another and my ticker to start working again. Takotsubo cardiomyopathy, a physical condition in which the heart muscle becomes suddenly weakened and the left ventricle changes shape, was identified in Japan in 1990 but the news hadn't reached me in Hampstead Garden Suburb in 1999. I reckon I had it, though. It was painful and frightening, but I thought I was being overdramatic and didn't bother telling anybody.

We blundered on somehow. The house seemed icy cold. I turned the heating to tropical and stayed deep-frozen. Echoing, enormous, it was a cavernous igloo. The girls kept falling over, slipping on the bathroom floor, tripping on the steps. It was as if their father's departure had upset their balance. They fought to keep upright. I fought to keep breathing. I was thinking in tiny survival increments. 'If I can get through the next four minutes, it will be five to twelve. If I can stay on track for the next three minutes, it will be two minutes to twelve.' I couldn't manage more.

I went to see a therapist. He looked like Topol and spoke in a middle-European accent. I found it difficult to sit still – at home I'd taken to pacing the floor, banging my hip on the wall, pacing back again and banging my hip on the other. He said comfortingly: 'You're nobody's *schmatta* [rag].' I handed over a hundred pounds. The next time I saw him, he heaved a heavy sigh and said: 'You're nobody's *schmatta*.' I parted with another hundred pounds and decided in future I'd be better off at the Temple Fortune branch of Hair on Broadway – at least I'd come away with a decent blow-dry.

Friends and family expressed disbelief. Cousin Beverley-on-both-sides said she'd watched only days before as the Good Doctor and I cuddled on the couch at her house, thinking how content a couple we were. No one said they had seen this coming. Reading the press coverage was hallucinatory. I didn't know what was happening in my own life yet journalists I'd never met or spoken to were writing all sorts of invented bilge as if it were fact. Most distressing of all was *Hello!* magazine, which used a mournful picture of me on its cover. I recognized it as a rare non-smiling shot, rejected from a *Telegraph* shoot eighteen months earlier. *Hello!* implied this was a photograph they had just shot to illustrate my current heartbreak and followed up with twenty pages of 'explanation' for the split, supplied by journalist Jan Shure. She occasionally commissioned articles from me for the *Jewish Chronicle* but didn't know me personally and knew nothing whatsoever about my private life.

I rang to ask how she could justify pretending to have intimate knowledge of my marriage. She didn't know me. She insisted all's fair in journalism and she was within her rights to sell a story about me. She told them I had worn the trousers in my marriage, after allowing fame to 'go to my head'. You know, Gentle Reader, I'd been careful not to do anything of the sort. Reading that article, illustrated with photographs of my wedding, portraying me as an egomaniac who had driven away a caring husband, was agony. I hated myself. Surely I must be repellent to send a perfectly delightful husband running for the hills.

55

Piers Morgan, thank you

REMEMBER ANSWERPHONES? OURS WAS IN the kitchen. The girls and I trailed in forlornly to see the light flashing. 'Hello, this is Sue Carroll, columnist at the *Daily Mirror*. Vanessa, your husband is having an affair. We'll be running pictures of the two of them in tomorrow's paper. Ring me if you want further information.'

The girls looked desolate. I hadn't been able to stop them hearing. 'Don't worry,' I said, trying to keep my voice steady. 'I'm sure this is a mistake. I'll go upstairs and have a word with her now. You two watch a bit of telly.'

Sue Carroll was brisk. 'He's hooked up with a junior doctor. She's called Rachel. She's only twenty-six. How old are you?'

'Thirty-seven.'

'Bloody hell! Who'd have thought you'd become the "older woman" at thirty-seven, Vanessa? They've been meeting in secret. Piers Morgan has had reporters tailing him ever since he left. Piers has a built-in bullshit detector. Anyway, buy a paper and you'll see what he left you for. No great shakes if you ask me.'

I rang the Good Doctor. He hadn't been taking my calls, but he picked up. 'Yes, Rachel and I are friends.'

'You've only been gone two weeks. We've been married nearly fifteen years. I've never heard you mention a Rachel.'

'We are friends. There's no more to be said. You may tell the girls she is just a good friend and that is that.'

'While we're on the phone, may I please talk to you about when you might want to see them?'

'No. I have nothing whatsoever to say to you.'

I went to calm the girls. 'It's fine, a misunderstanding. Daddy says the lady is just a friend. There's nothing to worry about. Don't cry, my darling sweethearts. Please don't cry. Everything is going to be all right.'

The phone rang. It was the Good Doctor. I went upstairs. There was that Dalek voice again. 'Yes, we are lovers. Yes, we are in a relationship.'

I have total recall of what happened next. I said: 'But I thought you were decent, honest and truthful.'

He replied so casually it sounded like a verbal shrug: 'Well, I'm not!'

Two days later Saskia heard two boys at school arguing over whether they'd rather get off with Saskia's mummy or Saskia's dad's girlfriend.

That was that. Divorce can be ultra speedy if you don't bog it down by wading through court proceedings. I had to pay the Good Doctor a shedload of money. If you're wondering why, if he left and was unfaithful, I had to pay him, you need to know that glittering among the United Kingdom's jewels of justice is 'no-fault divorce'. This means the baddie doesn't get punished. You can sleep with your wife's mother, sister and the entire school sixth-form choir and – unless there's proof of financial fraud – you still get half the marital assets. Lots of people don't know about 'no-fault divorce'. They assume if your husband does a bunk with a twenty-six-year-old he can't come for the family house, much less your earnings. They are wrong. I knew I'd have to sell the still horrifyingly mortgaged rock-star mansion. My divorce lawyer, Douglas Alexiou, warned me: 'Your husband might say he's going to let you have the house because you live in it with his daughters. Don't necessarily believe him. Many husbands say it. Not all of them mean it.' He never said it.

56

Divorce equals demotion

DIVORCE MEANS DEMOTION. YOU ARE instantly relegated to an undesirable sub-stratum, exiled to – you know what I'm going to say – the singles' table at barmitzvahs. Days after The Departure a pal I'd known since school asked me over for supper. If the date had fallen three days earlier, when I still had a husband, the table would have been laid in the dining room, children would have been packed off to bed, bottles of mediocre alcohol imbibed and sprightly conversation exchanged into the small hours. Husbandless, I didn't merit any of that. I was shown into the kitchen and served sausages and oven chips with the kids – no matter, I'd left mine behind with a baby-sitter. Her husband peeled off straight after shovelling the last banger into his gob. She pulled on a pair of Marigolds and gave her kitchen units a thorough once-over with a J-cloth. What was I meant to do? Applaud?

That's what I mean by demotion. A husband can be anything – boring, scrofulous, predatory, treacherous – but any husband is better than none. Without one you are just about good enough for a girls-only midweek coffee or to spectate at Wednesday kitchen cleaning. You are automatically disqualified from couples' dinners. Saturday nights are banned for singles. I may have been this woman's friend for twenty years but there was no way she'd get out the Royal Doulton for me until I'd dragged someone with a penis up the aisle.

There is a special hell reserved for women odious or idiotic enough to let a nice Jewish doctor slip through their fingers. You don't go to all the trouble of landing one only to let him trickle away with a mousy twenty-six-year-old. It's unforgivable. Women, even with a few notable exceptions, your closest friends, tighten ranks. They think you might slink round in a negligée and seduce their couldn't-fancy-them-pickled husbands. You wouldn't. You've just been driven into near suicide territory by your own husband cutting and running with his bit on the side. You wouldn't cause any other woman that kind of agony. Besides, how would it look in the *Daily Mail*?

I was more famous than ever. I'd morphed into perfect tabloid fodder. A TV downfall plus a divorce – just think of the dozens of Ds in the obligatory alliterative headline. The only missing D was DEAD and I was a cobweb's breadth away.

Friday-night dinner

Have I mentioned I wasn't allowed out on Friday nights till I was thirty-seven? Friday-night dinner, now immortalized in the eponymous sitcom, isn't optional. Other people dress up and get bladdered on Friday nights. Jewish people go to their grandma's and eat chicken livers, chicken soup and roast chicken prefaced by prayers over wine and bread to welcome the Sabbath. The men intone the blessings in Hebrew. The women make a blessing as they light two candles, then hover over the chicken, making sure the breast stays moist and the legs don't turn to charcoal. The women feed the men first, then the children, then the other women. By the time Grandma Sybil was ready to sit down and eat, Poppa Willie had polished off his first portion and was ready for seconds.

All my life, every single Friday night, including at Cambridge where I obeyed my parents' strict instructions to eat Shabbat dinner at the Jewish Society, I stood, inhaling the chicken aroma as a man – grandpa, father, husband – sang the words written in the sixteenth century by Shlomo Alkabetz: 'Lecha Dodi, Let us go my beloved to greet the bride, the Sabbath.' *Fiddler on the Roof* evokes the scene in the symbolic village of Anatevka as the women light the candles singing to their daughters.

For the first time since I was born there was no man to usher in the Sabbath. To my shame, I didn't know how to recite the prayers. Sitting in the dining room, eating a traditional meal when none of us had an appetite and we could barely swallow, was too much to ask of the girls or myself. Turning my back on nearly six thousand years of tradition, I put our food on trays and we ate in front of the television. I hurled my parenting standards into the wheelie bin.

The synagogue secretary phoned. 'Hello, this is Norrice Lea Synagogue.'

'Hello.'

'We hear your husband has left.' I hoped they might be ringing with some spiritual succour. Might the rabbi ask us for dinner on Friday night? Might they have a roving rent-a-cantor who could come round and deliver a benediction or two? 'So, we are correct? Your husband is no longer in residence?'

'Yes, unfortunately, that's right.'

'We are ringing to ask if you can still afford to pay your synagogue subscription fees.'

I could but I didn't. I dismembered myself – surely that's the right and proper verb – right then and there. I've never joined another. That means I don't pay the Jewish Burial Society so when I pop it there'll be nowhere consecrated and kosher to inter my bones. Oh, well, I'll just be cremated and have my ashes scattered over Brent Cross Shopping Centre. At least that way the girls will visit me a couple of times a week.

My father called to tell me the Good Doctor's departure was all my fault. 'What on earth did you think would happen if you let yourself get so fat?' He also insisted I change my will immediately. 'If you get run over this morning, the whole lot goes to him. That's no good. Go to a lawyer urgently.'

I wept so copiously over the female solicitor I soaked my last will and testament. She took pity on me. 'You seem so sad. Is there anything I can do to help?'

'Yes, please. I need a husband. I don't know how to survive without one.' Her uncle was a rabbi! What are the chances? She understood. She acted. The chairman of her uncle's synagogue was single. Should she give him my number?

The gentleman took me out for dinner in Hampstead. He was a braggart, tooting away on his own trumpet. I said almost nothing. I was still in love with my husband and horrified to be on a date with any man at all, particularly that puffed-up pill.

Never go on a dinner date

Note to self and all daters: dinner dates are dreadful. Limit the encounter to coffee or a bun. He smiled and said he'd had a wonderful time, so magnificent he'd like to see me again. NB Men often say that when you have been silent as the sodding Sphinx. There was, he had to tell me, just one tiny fly in the ointment. He had a partner. He'd been living with her for twenty years. He hadn't told the rabbi because the lovely lady in question wasn't Jewish. He thought it would be a tremendous idea to see me on the side, as his mistress. Should I have rung the rabbi and ratted the cheating chairman out? I was too busy trying to put one foot in front of the other.

Friends' horny husbands

A pathetic bunch of friends' and acquaintances' husbands did beat a straggly path to my door. How was I doing all alone in that huge house? Could they do something constructive to help? Change a plug? Massage my breasts? Have sex with me in the pool-house?

'No, thank you. And how is Susan/Nicola/Natasha/Anouska? Give her my love and tell her I'd love to see her.' The presence of the husbands underlined the absence of the wives. Divorce is death to friendships. Years later I wrote an article about this, and thousands of women (and a few chaps) wrote to describe their hurt as friendships disintegrated 'just' because a husband bit the dust. Most of the pals the Good Doctor and I 'shared' were mine. He'd only picked up a smattering, possibly because he'd been devoting his social energies to *les liaisons dangereuses*? Fresh tales of his conquests emerged every day.

Now the lid was off Pandora's box, people who'd kept his antics secret were suddenly eager to regale me with sights, sounds and smells.

Divorce means friends desert you

Friends I'd known since nursery school melted away. Apparently, the Good Doctor laid siege to them, paying personal house calls asking for 'help to manage this situation'. The divorce was all over the papers. It must have proved irresistible to be offered what felt like the inside scoop. For a while, I ploughed on, oblivious. I rang my lifelong girlfriends as I always had. It took me a while to realize the conversations ended not as before with a date to meet but with goodbye. They were avoiding me. They were otherwise occupied. We'd been childhood neighbours, school friends, antenatal-class partners. They'd carried in the required mourners' low chairs for my mother's *shiva*. I thought they were real ride-or-die friends, the fog, snow and ice kind, not showbusiness 'mwah-mwahs' you'd expect to be fair weather.

I can't have been much fun at that point. My repartee must have gone off the boil. You might have thought, though, they'd have understood my need to keep going over the cataclysmic details and possibly pumped out a few millilitres of human kindness. Ladies, you know who you are: A, L, D, F, E and a coven of others. I still don't know why you ditched me. Lord, it was a loss pile-on – lost mother, lost father, lost husband, lost friends. Add in lost career, lost money, lost house, lost self-esteem, lost will to live and you get the picture.

New pals bob up

A surprising thing happened. People I hadn't thought were friends stepped up. Mates from work came over and made dinner for me and the girls. Colleagues I'd imagined to be no more than professional contacts seemed to grasp the raw misery of surviving each twenty-four hours and were prepared to pitch in. New friends miraculously appeared. I thank the Lord for them. They know who they are, and I will – and have – criss-crossed the earth for them. They rescued me.

57

A quick cautionary tale

Old School Pal

BEFORE THE BREAK-UP I'D BUMPED into a bloke I hadn't seen since school. It was no big deal. He was no big deal. Out of politeness I obliged when he suggested swapping phone numbers. Learning of my woes, like the rest of the nation, he turned up with a bunch of grapes.

Here's another horrible side effect of fame: when the shit hits the proverbial all sorts of creeps who happen to have your number – or, these days, can winkle you out on social media – slither nosily out of the woodwork. Their aim is to get the lowdown on whatever ghastliness is being peddled about you. If they can prise some nugget of insider info out of you, hey presto – lashings of cachet. The trick, once you wise up, is to reveal absolutely nothing. Get off the blower without donating a crumb of gossip. I've also – no shame here – done the occasional Coleen Rooney and planted droplets of false information. It's a fun way to find out who can be trusted. 'How am I? Well, don't tell a soul, but I'm so much better since I ripped out the Good Doctor's office and had it replaced with a personal hairdressing salon and massage area.' Sit tight and watch the story fly.

Back to Old School Pal. If I'd been able to like anything, I'd have quite liked seeing him. He remembered me in my schoolgirl prime as Vanessa the Undresser. He wrote poetry and plays. Friends Reunited didn't start till June 2000. This was November 1999, but the effect was identical. In Old School Pal's company, I was myself pre-marriage, pre-children, pre-fame, pre-downfall and divorce. It was a blessed relief. There was zero chemistry. We'd always been platonic friends. (I had quite fancied his older brother.) OSP invited me to cool stuff I'd never have seen when I was married: an off-off-West End musical, a Bollywood film in Southall.

The few hours with him were a diluted approximation of fun in weeks polluted by lawyers, accountants and pernicious faxes from the Departed Doctor (one notably setting out his plans for division of our air miles!). He joked. I laughed, which was a great deal better than my perpetual weeping. I couldn't eat, even at the offbeat intriguing restaurants he picked, but I could watch him eat and talk and enjoy it.

One night at his cat-friendly flat, he made a move. He dived headlong upon me from above. I flinched. It was a reflex. I couldn't hide or help it. He noticed and was obviously hurt, angry or both. He dropped me home and his coldness made it clear that was the end. I couldn't face losing him and our Saturday-night excursions. I didn't have the emotional ballast to contemplate another loss. I was sure he'd be careful with me. We'd known each other twenty-three years. He wasn't some stage-door Johnny. He'd been my escort to my parents' twentieth wedding-anniversary party. He was Jewish with literary pretensions, just like my father. I could, I thought, trust him to be kind.

So, I did what many women have done before and since – persuaded myself I could manage to fancy him if I just tried hard enough. Alcohol – and, as you know, I barely touch it, much preferring chocolate mousse – played a significant part. I managed to alleviate the awkwardness of the 'Yikes! No! Don't touch me!' flinch by pitching up at his office with a *wurscht* (kosher salami). Yes, there were phallic jokes to go with it. Yes, I was humorously saying: 'I was just shy. It wasn't personal. I would happily lick your sausage.' It wasn't true but I was determined to give it my best shot. I knocked back a few glasses of white wine. He made a complicated dinner. He'd also bought new dining chairs at auction specially and chosen an array of films on video. Being cared for and about was balm to my serrated soul. I'll draw a ladylike veil over the sex part but lift it gently to report it was better than satisfactory and, having partaken, I was more than happy to partake again, and again.

OSP was love-bombing me. I didn't realize it, of course. The term hadn't been coined yet. He bought me gifts: tiny crystal *chotchkees* (trinkets), a gold necklace engraved with *The Tempest*'s 'We are such stuff as dreams are made on', glass earrings from the British Museum shop. He took me and the girls to his parents' for – you guessed it – Friday-night dinner. Everything about their abundant hospitality, fragrant chicken and welcoming smiles was achingly familiar, like being ushered back to a warm, safe home from which you've been banished.

Here's where I went wrong. The term 'red flag' hadn't been invented, but whatever they were called, I didn't spot them, and I should have seen a flotilla of them coming a mile off. Old School Pal had never been married or lived with anyone. He'd had one fleeting engagement decades before but should have had 'commitment-phobe' tattooed on his forehead. All the things I thought were pluses – he's single, he's available, he's baggage-free – ought to have rattled my antennae. I did the hasty, the ridiculous, the abominably stupid – I fell in love.

He seemed obsessed. He drove me up verdant Downshire Hill in Hampstead and said: 'This is where we'll live when we're married.' He talked about babies. Would I? Yes. Could I? Yes, I'd just turned thirty-eight. He took me to his aunt and uncle's anniversary shindig and introduced me to his family. He whisked the girls and me to the *Sing-A-Long-A Sound of Music*. If we were meant to meet at eight, he'd arrive at seven because he 'couldn't wait' to see me. The pavement sparkled. I couldn't stop singing, mostly Dinah Washington's 'Rockin' Good Way (To Mess Around And Fall In Love).'

Fifteen weeks later, Old School Pal pulled the plug. He'd been unnervingly hot and cold, swooping in and withdrawing in an unsettling way. One second, he couldn't get enough of me, the next he decreed I mustn't make any contact with him at all. I was distressed and confused. I began to chain-read self-help books. Was he behaving like an unpredictable potentate because he was from Mars? What the fuck? Of course I should have told him to sling his hook. He was only Old School Pal. I'd had to try my best to persuade myself to fancy him. He lived in a flat with cats.

He came for – yes, you guessed it – Friday-night dinner. He asked for second helpings of every dish I lovingly carried to the table. Flocks of chickens died to fill his rapacious belly. I was relieved he was there, pleased he liked my cooking, dashing about with a gravy boat in maximum Jewish-geisha mode. When the girls went to bed he said: 'I'd like to bring this to an end. You are just too needy.'

Needy? You think so? I'm a few months into a bloody divorce. I've confided every ounce of my shock and misery. Now, you're accusing me of being exactly what it's glaringly obvious I am and have never tried to hide – in need of love, constancy and loyalty.

A monumental mistake

OSP was a monumental mistake. You probably spotted some of my errors. Worst of all was involving my wounded, vulnerable daughters in a fledgling relationship with a trifler. Obviously, I didn't know he'd chew me up and spit me out so quickly. We'd known each other more than two decades and I'd hoped that meant the relationship would be stable and lasting. I should have kept the girls at a distance from him, not presented the three of us as a package. I ought to have seen him outside the house away from my children. I clearly see that now. At the time, though, they were so newly hurt I didn't want to be separated from them, even for an evening. I had the maternal desire to hold and hug them and gather them to safety beneath my wings. I believed I was helping them by including them in what I hoped would be a solid relationship. I didn't want them to feel excluded and I needed him to know they were the most important part of me.

Yet again I was wrong. In trying to protect and enfold my daughters, I unwittingly hurt them. Even thinking about it twenty-four years later, I'm sad and overwhelmingly sorry. However heartfelt my apology, it will never be enough. I hope they weren't irrevocably scarred by my lousy choices. I wish I had been perceptive enough to sense I was putting them in harm's way, not shielding them from pain. I am a loving mother. I couldn't love my girls more ardently. Yet I have been a less than flawless mother. And, lest you ask, knowing Philip Larkin was relaxed about parents fucking up their kids doesn't make me feel any better about it. I think I should have tried harder.

I was left with a windowsill full of horrible crystal figurines – I could admit they were ghastly now he'd given me the Big E. I gave them to one of Allegra's friends as a batmitzvah present. They were just right for a twelve-year-old. I hurled into the rubbish the expensively framed newspaper articles about my 'new love'. They'd only been up on the wall for about three weeks.

58

On the bright side, you're gaunt

LOOK ON THE BRIGHT SIDE. I'd lost a truckload of weight. Better not fuck it up, Vanessa. Don't screw up the one thing you've got going for you. Don't eat. Anything. Ever again.

I carried on training with the Angel of the North seven days a week – no exaggeration, Sundays included. We worked out in silence. He tried playing music. I couldn't bear to hear it. He tried uplifting pep talks. I was crying too hard to hear a motivational syllable. I rowed in wretched silence, bounced on a mini trampoline enmired in misery, ran on the running machine. I tore up and down three flights of stairs. I squatted furiously. I did that hideous thing where you bend your knees, park your hands on a bench and pump yourself up and down until, weeks in, you see the bump of a bicep. I flogged and slogged, sweated and strained. The highlight was pulling on boxing gloves and beating the crap out of the Angel. Obviously, he could have flattened me with one thump, but he let me pretend to think I'd pulverized him.

I hated every second. Physical jerks are alien to my sedentary, sip-iced-coffee-with-a-freshly-squeezed-Balzac-in-one-hand-and-a-taramasalata-bap-in-the-other nature. I didn't have that option. I was too anguished to read, too fraught to recline, too heartbroken to . . . eat!

What did I just write? Yes. You read it correctly. The unimaginable had happened. A dead mother didn't do it. A disappearing husband did. My appetite was annihilated. I think I realized food wouldn't cut it. Sticky toffee pudding couldn't erase the sight of my daughters' anguished faces as their father drove away. If I knew food wouldn't alleviate my suffering, what was the point? I grew thinner. A paparazzo snapped me coming out of the chemist. The picture of me looking lost and listless in an anorak was captioned 'Gaunt! Vanessa's friends made anxious by star's rapid weight loss'. Only my remaining pals weren't anxious at all. They were overjoyed. 'You're slim. No, you're not even slim. You're verging on thin. You can fit into a Karen Millen size twelve embellished dress and tiny cardie combo. This is brilliant. OK, you can't breathe, sleep or see straight. Tears are dripping off your chin and you are ferrying yourself and the girls from therapist to therapist just to survive each day. You're thin! You look terrific. Well, apart from the shadows under your eyes, ghostly pallor and shaking. You're nearly skinny. Keep going, girl. You'll get there. And when you do, you'll find a fresher, better husband who'll stick around. Don't give up.'

If I Can, You Can

Universal got in touch. How did I feel about doing a fitness and weight-loss video? I felt nothing much except 'Where do I sign?' A video would give me something to focus on. Training was, apart from the girls, the only fixed point in my life. The divorce was set to cost me a fortune I hadn't even earned yet. I might as well make a few quid. As my father asked rhetorically most days: 'Who's going to give you a pound?' Universal offered quite a few. They signed up the Angel of the North and he got to work devising a routine to make me look good and reflect our real daily training programme.

The video was called *If I Can, You Can*. It was filmed over two days at my house. A catering van arrived in my front garden. The girls were in clover. We could sit in our own drive and tuck into three-course meals in the 'restaurant' Winnebago. Or they could. I still wasn't eating and the pressure of appearing at ease while bouncing about toting weights and lifting dumbbells wasn't exactly an aperitif. The Angel of the North and I hurled ourselves around good-naturedly in an assortment of skintight Lycra, as director Stuart Watts yelled encouragement. I had a headful of waist-length extensions fused to my roots by hairdresser Errol Douglas. The idea was that the Angel of the North's dreadlocks and my abundant ponytail would swing through the air in ebony and ivory-esque unison. We repeated each routine about twenty times so the cameras could capture every detail. I was depleted. He was energized. By the end of filming, I'd lost another half-stone. I was a slim size ten. I hadn't been so minimal since my sixteenth birthday.

The Angel of the North

If I Can, You Can sold up a storm. Angel and I squatted and did star jumps at branches of Boots, in the aisles at Asda Park Royal and in WHSmith up and down the country. We took the show on the road touring theatres in a live *If I Can, You Can* extravaganza in which we whirled around the stage on scooters, romped through a fitness rap, inveigled innocent audience members on stage to join us in vigorous physical jerks and somewhere along the way fell for one another. I was happy. I was skinny. I was the co-presenter of the most successful fitness video in the country – for a short time, admittedly, until some other skinny Minnie's video knocked us off the top slot. Angel was still training me. I was still leaping about, tearing up and down stairs, bouncing my boobs off on the trampoline, but something had changed. Instead of being so suffused with sadness I could barely keep down water, I was hungry.

Caribbean food was as beguiling as Angel's company. I was wolfing down patties, bun and cheese, akee and saltfish, festival dumplings and fried plantain with love-is-in-the-air gusto. For the first time ever, I was famously thin, but eagle-eyed observers (another name for paparazzi) were starting to notice I wasn't quite as skeletal as in the picture on the *If I Can, You Can* cover.

Imagine bonking as physically potent a man as the Angel of the North after a lifetime of canoodling with bookish, poorly coordinated Freud-fixated doctors and accountants. Lord, it was a revelation! He was an Exocet missile of turbulent testosterone, and I was intoxicated. He was also an enthralling spinner of stories, practising Buddhist, political pundit, driver of convertibles. His adorable four-year-old daughter, Summer Pearl, brought sugar and spice to our lives, and his exuberant extensive Jamaican-Mancunian family welcomed the girls and me to their almost weekly weddings, confirmations, anniversaries and wall-to-wall shindigs with open arms and vats of steaming curry goat. We wore out the M1 to Manchester visiting Angel's hundreds of relatives, my uncle Clive and dear pal Sara. Angel wasn't keen on rocking up red carpets. He loathed the limelight, much preferring to tread his own path privately. He hated it when journalists turned up at his parents' house in Old Trafford sniffing about for tittle-tattle. He wasn't altogether pleased about slotting into the husband gap either. I longed for till death us do part. He wanted to 'walk street', preferably in the direction of a party. We electrified each other till we were millimetres from short-circuiting in a megawatt shock. Could this febrile union last?

59

Celebrity Big Brother

BY 2001 MY NOTORIETY WAS undeniable, but my career had evaporated. Enter Richard Curtis. Yes, Richard *Four Weddings and a Funeral*, *Notting Hill*, *Vicar of Dibley*, creator of Comic Relief, venerated by all as a present-day saint Curtis.

'Vanessa, Richard Curtis here. I hope you don't mind my contacting you like this. I need to ask you a favour. You've heard of *Big Brother*?' Of course I had. Alongside the rest of the nation, I'd watched the first ever series the year before and booed 'Nasty' Nick Bateman from my sofa. 'Well, we're going to make the first *Celebrity Big Brother* for Comic Relief. There's no fee. It's all for charity. You'd just go into the *BB* house with a bunch of other celebs, can't tell you who. You caper about for a week at most. All the money raised by voting goes to Comic Relief. It'll be shown half on BBC1, half on Channel 4. Lord knows if anyone will watch the thing. Anyway, are you in?'

No one says no to Richard Curtis. Of course I was in. How hard could it be? I imagined the experience would be a bit *Blue Peter*, a bit Lenny Henry playing a good-natured game of street football with some smiling kids in Soweto, except somewhere in East London.

In March 2001 there was no such thing as reality-TV fatigue.

The genre was new and intriguing. Serious people chaired intense debates about voyeurism, George Orwell and whether Shakespeare would have been a *Big Brother* housemate if he'd had the chance. It's hard to believe it, but twenty-three years ago, no celebrity had ever been seen on the small screen without hours in Hair and Make-up, best bibs and tuckers and a collection of rehearsed, carefully curated anecdotes to deliver to a faux-amazed host. Household names wandering about in their pyjamas buttering pieces of toast and arguing about cleaning the loo was not yet a thing.

We were about to change all that. I say 'we' but on the morning of March 2001 I had no inkling who my housemates were going to be. A special bright red *Big Brother* suitcase was delivered to my house and a car arrived to transport me to the mystery location. The girls were going to be allowed to stand in the crowd of relatives and friends to wave me off from a distance but would be collected later. We were forbidden to travel in the same vehicle. Then and there my resolve started to crumble. I felt miserable being driven away from my children. I'd kept them close – probably too close – since the divorce just a year earlier. I would be incarcerated in an unknown place with no phone contact and no way of finding out how they were doing. It suddenly seemed an appalling idea. I was disgorged and taken straight to solitary confinement by silent researchers.

You must have gathered by now I don't flourish alone. There was nothing to read in the cell-like room, no TV to watch, no phone and no one to talk to. I had far too much time to think. After a few hours' staring at the wall, realizing I had no notion what torment I'd let myself in for, I was already unravelling at the seams. Finally, I was led, wearing a warm coat and hat and carrying my case – the days of arriving at *Big Brother* in sparkling evening gowns and emerging, shimmering, from chauffeur-driven limousines hadn't yet begun – to a concrete concourse. Presenter Anthea Turner, comedian Jack Dee, boxer Chris Eubank, soap actress Claire Sweeney and Boyzone band member Keith Duffy stood shivering in unflattering outerwear looking apprehensive. My girls were too far away to cuddle. I know it had been only a matter of hours since I'd seen them, but everything was so hyped up, I felt sure they'd lost weight and shrivelled. They looked like starving orphans stranded in the forest, in dire need of their mother.

The set dressers had had a field day. There was a tangle of barbed wire, huge howling dogs slavering on metal leashes and a gigantic castle door to bolt behind us with a terrifying clang. I was a seasoned telly professional. I ought to have been immune to the scary effects designers can create with a glue gun, staples and plastic Halloween props. I should have been able to ignore the snarling dogs and pantomime chamber-of-horrors kit and smiled through the lot. Instead, I was unnerved and trembling. The door crashed shut. We were imprisoned, at the mercy of one another and the innumerable cameras trained upon us from every angle.

Yes, I'd known we'd be filmed twenty-four hours a day everywhere except in the bathroom. Knowing something is nothing like living through it. The sensation of remote-controlled cameras twisting round to capture you rolling on deodorant defies description. You feel violated and incredibly important all at the same time. An irregular thing happened seconds after I entered the house. Through an open door I saw a dozen immaculate fine white linen shirts on hangers being delivered to Chris Eubank. What on earth? After all the fuss about the regulation suitcase and not smuggling in a single extra item, Eubank was already circumventing the rules with a special delivery.

I didn't protest. My aim, shared with all my fellow housemates, was to fit in, be liked by the public and stay in the house as long as possible. Quickly Jack Dee became leader. In 2001 it seemed perfectly acceptable that the middle-class white man instantly took charge. He announced he'd make the lunch and begin by baking bread. We would act as his handmaidens and disciples. I knuckled under and began peeling, dicing and washing up as if my life depended upon it. Candidly, my life (in the house) did depend upon it. If I was likeable and down-to-earth, viewers would vote to keep me in. If I was abrasive, lazy or selfish they might evict me with humiliating haste. We were all on our best like-me-please-like-me behaviour, except Chris. Eubank didn't give a damn. He put on a caftan, stuck a monocle in one eye and buzzed like an infuriating mosquito around the kitchen on his skateboard. How come he had a skateboard? No one else was permitted such a thing. Every time he spotted a mirror – there were dozens, cameras concealed behind them – he'd stop and talk to his reflection. He fancied one of the female voices of *Big Brother* and spent ages trying to chat her up. Unlike the rest of us, he didn't wash up, offer to make tea or muck in. He ate the food we'd prepared, slurped the tea and refused to lift a finger. I couldn't help admiring his chutzpah.

We sailed along as amicably as six people who don't know each other but are energetically pretending to bond can until Big Brother announced we'd be called into the Diary Room to nominate one another for eviction. It wasn't a shock. I knew the rules of the game. I hadn't anticipated how horrible the nominating process would feel. I didn't want to single out any of my newfound 'pals'. I didn't want viewers to think I was a hypocrite. I didn't like being two-faced in private, let alone on camera. I chose Jack and Chris. All six of us loathed nominating. It wasn't fun or light-hearted. The atmosphere deteriorated dramatically.

An hour later Big Brother announced Anthea Turner and Chris Eubank were up for eviction. Anthea burst into tears. I understood completely. It was a popularity contest. She didn't want to be disliked. Let's face it, no one does. She was scared to be first out, upset her housemates (not me, thank Heaven) had chosen her for eviction. She was sensitive and vulnerable – she'd recently survived a brutal roasting in the media – and she couldn't hide her discomfort.

It was awful. A forty-year-old woman was crying her heart out, cameras capturing every tear. It sure as hell didn't feel as if we were playing a game. Chris said he refused to nominate again. He couldn't stomach it. Jack Dee wondered if we could just pick matchsticks to avoid hurting feelings. Anthea kept on crying. It was hard to comfort her. We couldn't say, 'It doesn't matter who goes first,' because it *did* matter, very much indeed. We'd only been in the sodding house since lunchtime and none of us had bargained for any of this.

Big Brother called me to the Diary Room. 'Vanessa,' said a producer through the sound system, 'we need your help. Housemates are refusing to nominate, and we can't have that. If they don't play the game properly donations to Comic Relief will be dead in the water. You've got to help us make sure they don't sabotage the show's success. Please go back and explain this is a game, with rules, and they absolutely must abide by them.'

Enlisting me, behind my housemates' backs, to persuade them to stick to the agenda, was a glaring rule-break. Nevertheless, I obediently went back and did my best to bring everyone back on track.

Keith wanted to spend 95 per cent of the food budget on cider. Anthea revealed the optimum way of changing a duvet cover. Claire sang snippets of *Chicago* well enough to bag a part in the show later that year. Jack Dee pretended to 'break out' of the house and turned up with mud smeared all over his face. There was no mud. It was a put-up job. Chris Eubank was first to be evicted and took his punishment stoically. Anthea and I wept in one another's arms when he left. I know. I know. It was an absurd overreaction but don't judge till you've been a guest of Big Brother.

It was obvious I'd be nominated for eviction next. It would have been downright cruelty to choose Anthea. Keith and Claire were young and, though Claire's soap *Brookside* and Keith's band Boyzone were hugely popular, as individuals they were still relatively unknown. They deserved to escape the chop. Jack Dee's name and mine were announced. I knew I'd be evicted. Jack was playing a blinder. 'Send me home. I hate it here,' he whined. Reverse psychology meant the British public would keep him in just to spite him. I'd be going home, but not for thirty-six hours. This was Sunday morning and eviction wasn't scheduled till ten p.m. on Monday night. I should have been able to hang in there. What's thirty-six hours in the scheme of things? I should have been able to keep a lid on my emotions but, just like Anthea, I couldn't.

Rapidly I spiralled into an emotional tailspin, and the fact that I happened to be wearing a leopard-print dressing gown and black sunglasses indoors at the time added to the spectacle. We'd been given chalk. I'd noticed the dining table was made of blackboard. I decided I'd write some appropriate words just to give myself something interesting to read later. I wrote: IMMOLATED, ISOLATED, INCARCERATED, IMMURED, DEFENESTRATED, LAMPOONED, DISTRAUGHT, FRAUGHT, ABASHED, DISCOMBOBULATED and a whole lot of cries for help.

Mid-scrawl, Big Brother demanded the chalk back. I told Big Brother to 'Fuck off!' Note: I was the first housemate in the world – *Big Brother* is a global phenomenon – to tell Big Bro to fuck himself.

I was summoned to the Diary Room for a dressing-down. I'd had enough. 'There is no Big Brother,' I said. 'You are a researcher who has drawn the short straw and had to work on Sunday morning. Please tell your bosses I want to go home. Now! I'm not being paid. I miss my children. I know I'll be out tomorrow night at ten but I don't want to wait that long. I want to go home. Right now.

'Please. I just want to go home.' My face was red and blotchy. Tears splashed out from under the dark glasses. I was so exposed I turned the revolving chair away from the camera so only the back could be seen. Ever since programme-makers Endemol have ensured all Diary Room chairs are anchored so housemates can't spin round and hide their faces.

My daughters say I was a reality-TV pioneer. I fell apart, rivulets of tears splashing into my cleavage. The nation was aghast and fascinated. Celebrities didn't behave like that on TV. Since then, an on-air meltdown is practically obligatory. Contestants are expected to unravel, rend their garments and demand through sliding screeds of snot to go home immediately. These days, therapists are on standby to dash to the rescue. In 2001 there were none. The voice of Big Brother was brutal: 'If you walk, Vanessa, you'll be the most hated woman in Britain.'

'Mental health' hadn't been invented.

Imagine saying that to a sobbing celeb now? You'd be sued, hung out to dry and cancelled. In 2001 Big Brother could bully me in any way he chose, and I hung my head in existential despair and took it. Of course I couldn't walk. I didn't want to be hated. I couldn't understand why I'd be despised for going home when I was about to be told to go home anyway, but I was in no condition to argue. Eventually I pulled myself together and wrote my *Daily Express* column on an ancient manual typewriter that Big Brother inflicted on me as a hilarious gag.

When, as I'd predicted, I was evicted, just three nights and four days after I'd been interned, I was disoriented – until, five minutes later, I saw my girls and the Angel of the North, slouched on my own sofa at home and felt normal again.

Channel 4 had to shell out for a bodyguard for me for the next six weeks. Why? To protect me from cuddles. Fifteen million people had watched the show and it felt as if every single one wanted to hand me a Kleenex and give me a hug. The postman delivered more than seven thousand letters, some addressed to 'Vanessa, Big Brother'. Some people kindly offered me and the girls a cosy billet in their back bedroom.

Twenty-three years later, the shock of me being real on reality TV endures. Maybe real is too much for viewers to cope with? I reckon I'm still asked what the fuck I was doing writing on the table on average three times a day. I used to say, 'At least it means there's always something to chat about at dinner parties,' but no one gives dinner parties any more.

60

Oh yes I did!

I WAS EATING AGAIN. OVEREATING. If only I'd done what I told everyone else to do on my eminently sensible video, I'd have kept the pounds off. You know me, Gentle Reader. Food, as usual, had nothing to do with physical hunger, everything to do with emotional hunger and was way out of control. It's horrible being fat and famous. Being newly skinny and famous is far more fun. Beware, though, being a famously skinny person gradually getting fatter. It's as if you're deliberately setting a trap for the media. They want to outwit you. Their cameras are poised to catch rolls of the fat you said you'd shed pouring out of your swimsuit bottoms. It's a money shot. 'Oh, no, Vanessa', 'Worried friends think Vanessa's drinking custard', 'Feltz eating meltz'. Obviously, I was desperate not to put the pounds back on. I was caught in a perniciously public yo-yo dieting cycle. I'd lose the weight time and time again, succeeding triumphantly in 2002, 2004, 2007 and 2009. Each time it was harder to shed the surplus. Dieting had to be more extreme, exercising more violent.

Yo-yo dieting

Going to bed hungry, waking up hungry and staring at an almost empty plate at mealtimes was essential. Months of privation crawled by. The misery paid off. Slowly I'd get slim, slimmer, slimmest. The paparazzi would snap it. Magazines would commission features about it. Eve Pollard, Claudia Winkleman's formidable mother, even persuaded me into a high-thigh swimming costume for one saucy set taken by John Swannell.

Yo-yoing sucks. Stay slim and I was congratulated. Strangers told me how gorgeous I looked – so much better than before – and companies coughed up cash. Fat, I was a national disgrace, an embarrassment to my family, an abomination to be outed and vilified. My health, prospects and earning capacity were all better if I was thinner. It's hard to grasp why, if that was so flipping obvious, I just couldn't keep the blessed weight off.

This is what happened with sickening inevitability. Elated at my success, relieved to stroll into 'normal' shops and slip gracefully into size twelve jeans, I'd celebrate my stellar achievement with a slice of strudel. The texture of pastry and fragrant apple on my tongue was so beguiling I'd have a piece of cheesecake the next day, and a side order of chips the one after. In no time at all, the weight piled straight back on and then some. Oprah Winfrey said the day after she wheeled the trolley full of bags of lard on stage trumpeting her weight loss to an awe-stricken world, she'd already started eating normally and gained four pounds. With each yo-yo, loss took longer, and gain happened virtually overnight. The paparazzi would out me. I'd be enveloped in shame. My arse would be snapped, stuck on magazine covers and its acreage shared with the nation. A woman in Temple Fortune Waitrose said: 'My God! Look at you. It's easy to see why your husband left you.' It wasn't fun.

Heineken, Jimmy Savile and a lion from London Zoo

2000. Would you like to do a commercial? You betcha. It's for Heineken. Lord! I've never had a sip of the stuff. Is it the one that refreshes the parts other beers cannot reach? Yes. Well, I still haven't tasted it. Does that matter? Not in the least. Is there anything else I need to know? Yes. You'll be in a series of four advertisements alongside other British telly stars. The shtick will be 'If you don't buy more Heineken, we'll keep on showing this commercial.' In the grand finale you'll all be eaten by a lion.

Gentle Reader, CGI hadn't reached its stride in 2000: a lion meant a real, live, roaring *Daktari*, *Born Free* lion. We filmed for three days at Pinewood Studios. The King of the Jungle arrived straight from London Zoo and growled in a cage in the wings, taken for occasional strolls on a chain held by – I thought – a rather timid-looking keeper.

The constellation of fellow 'stars' twinkling about the Pinewood firmament were: national treasure, friend to the royal family, host of *Jim'll Fix It* and Stoke Mandeville Hospital fundraiser Sir Jimmy Savile; former footballer turned beloved pundit Jimmy Hill; magician, 'I like it – not a lot – but I like it', Paul Daniels and his wife/sawing-in-half prop the lovely Debbie McGee; nightclub owner/fabled fornicator Peter Stringfellow; socialite Tamara Beckwith; and Lisa Riley from *Emmerdale*. It was never spelled out to us that the premise for the whole shebang was that we were all so irritating the audience would neck extra Heineken just for the pleasure of not having to watch us in the ads.

Jimmy Savile and Paul Daniels locked horns on sight. I enjoyed standing on the sidelines watching their cacophonous clash of egos. Both vied with the director to be senior alpha male, cock of the walk, big cheese and boss the rest of us around unmercifully.

I could pretend I saw through Sir Jimmy's palsy-walsy act to the evil throbbing just below the cigar/tracksuit disguise, but I didn't. Something repellent about him – dead fishy eyes, whiff of BO emanating from the tracksuit – stopped me saying more than 'Morning' but I'd be lying if I said I realized we were singing the Carpenters' 'Close To You' with a predatory and pernicious paedophile.

Advert number three featured me and Peter Stringfellow, suspended in mid-air. I was seated, leaning on a crescent moon, strumming a lyre, dressed as an angel. Peter was hanging upright dressed as Cupid. The director would shout, 'Cue Vanessa, cue Peter,' and from our dizzy height we'd start singing the sentimental lyrics.

Just as we built up a head of jarringly out-of-tune steam, Peter would yell: 'My balls. Ouch! Agony! They're chafing! Get me down.'

The director would shriek: 'Cut! Lower Peter. Wardrobe, free Peter's balls. Peter, let them swing about a bit. OK. Have you got the feeling back in your bollocks? You have? You're sure? OK, Wardrobe, get Peter back in the harness. Winch Peter up. Everybody quiet. Take two!'

Peter and I would start warbling romantically. We were trying to give 'Close To You' every ounce of romance we could muster. We got about halfway and Peter emitted an agonized yelp. 'My balls! They've gone numb. You're hanging me by my fucking testicles. For God's sake, get me the fuck down.'

The director went into freeing-Peter's-balls mode all over again. 'Cut! Lower Peter. Wardrobe, release Peter's balls. Peter, let me know when the feeling comes back. OK. Relax. Just swing them freely about a bit. Get some breeze on them. Right, all better. Winch Peter up.'

If I tell you Peter's balls were captured and released seventeen times before he managed to sing through the pain and deliver us from the sound stage, you'll understand why the director's beard turned white with strain. And why Lisa Riley going AWOL, having hijacked a stray golf cart, passed almost unnoticed.

Panto

Oh yes I did! Of course I did. Months after *Celebrity Big Brother*, I signed up to panto in Woking. I played the Genie of the Ring in *Aladdin*, starring Gary Wilmot and John 'I'm Free' Inman as Widow Wanda Twankey. I sailed into rehearsal on day one, imagining myself a gracious adornment to the proceedings. Wrong! There's an inviolable panto pecking order and mere TV personalities are lowest of the low. We're barely tolerated. We're there to get bums on seats, but we are not to be accorded even the minutest modicum of respect. We are not thespians. We are un-RADA-trained interlopers. We were not born in a trunk in the dressing room of the Plowright Theatre in Scunthorpe. We never carried a spear thirty yards to the left of Sir John Gielgud. We don't know stage left from stage right. We don't use the phrase 'off book' to mean 'I know my part by heart and don't need to lug the script around with me.' We don't call the proscenium arch the 'pross arch'. We are *persona non grata*, given the scuzziest dressing room and sidelined at the Christmas party.

Each performance began with me and Gary Wilmot underneath the stage waiting to leap out of a dramatically opening trapdoor. Sometimes we'd be waiting down there for twenty minutes. I tried to strike up conversation. Silence. I tried again, louder. Silence. I couldn't work out why he was shtum. Was he deaf? Was he cross? WTF? It turned out I'd infringed theatrical etiquette. As a mere TV person, it did not behove me to address the star first. Think Her late Majesty. It didn't do to stick out your mitt to shake the royal paw. You had to wait to be invited. The same protocol applied underneath the cobwebby stage of the New Victoria Theatre, Woking. I wasn't meant to speak until spoken to.

Aladdin 2001. Princess Badroulbadour wept noisily all day every day in her dressing room. The poor darling's heart had been cruelly shattered by an erring swain. John Inman was dressed – as a windmill, Concorde, washing-machine, the Statue of Liberty – by his inamorato Ron. Sometimes Ron would be indisposed, in a state of exaggerated merriment or asleep in the wings. Then there'd be a kick bollocks scramble as Aladdin's dresser disappeared to dress John and, like a collapsing pack of cards, everyone in the cast's dresser buggered off to dress someone else, leaving the lesser-spotted ensemble stark naked. One of the chorus was a terrifically efficient drug-dealer.

Aladdin took a million pounds at the box office within days of opening. Allegra and Saskia were enchanted. Saskia, eleven, sat through dozens of performances. She was word perfect and occasionally prompted from the stalls. She still quotes entire monologues verbatim and, if pushed, will give a mesmerizing montage of the songs. My father walked out of the first night in disgust, hissing the production was 'too commercialized'. It wasn't. There were no commercial tie-ups whatsoever. It was a traditional, old-fashioned panto. My father enjoyed turning up to things I was involved with and being too repulsed by something I'd done or failed to do to suffer another second. By 2001 I should have learned my lesson and left him off the guest list but, like millions of praise-deprived adult children, I couldn't help wishing for his approval and trying my hardest to earn it.

61

Saskia's batmitzvah

THE DAY AFTER HER FATHER left, Saskia crept into my bed in the middle of the night. She was ten and, bless her heart, worried about her batmitzvah. Although it was two years away, she feared it wouldn't happen without her dad and was anxious we'd have to move house and her celebration would be different from her big sister's. I promised her we'd stay right there in our beautiful home: she had nothing to worry about. Her celebration would be as wonderful in every way as Allegra's had been. Then I started worrying.

I'm still relieved and happy to report Saskia did have the glorious day she deserved. The sun shone. She looked a picture in palest blue. We had a dancing luncheon so warm and brimming with love for her, our guests, children and adults alike, climbed up on their chairs, linked arms and swayed along with the *klezmer* band. Her *d'var torah* (learned speech) was so beguiling Dame Esther Rantzen sobbed. In the evening, we had a disco for all her friends, featuring an ice-cream van and a pizza wagon. She was the most adorable, empathetic, humorous, gorgeous little girl and she has grown into an adorable, engaging and remarkably creative teacher and child therapist. I wish I could throw a party celebrating her effervescent personality every week.

Farewell rock-star mansion

I'd been clinging to the rock-star mansion by my nail extensions. Batmitzvah accomplished, I had no choice but to sell. The Good Doctor demanded his pound of flesh, and I couldn't stall any longer.

Gentle Reader, you know perfectly well I never wanted the house in the first place so you might wonder why it turned out to be such an almighty wrench to get shot of it. I suppose it's because it was our home. The girls and I were radically unsettled by the divorce. I'd lost my job. It wasn't a great time to pack up everything familiar. I tried and failed to give away enormous pieces of barely-sat-on furniture. No one had room for it.

Fetching up somewhere none of us wanted to be, surrounded by cardboard boxes, wasn't fun. I told you adjusting to majestic splendour took seconds: gearing up was effortless. Adjusting downwards was harder. The new house felt smaller, pokier, darker, because it was. The late film director Michael Winner told me: 'I've been to your house – too many small rooms.' In fact, there weren't that many: four small bedrooms, a kitchen, living room and reception room. The place is gingerbread cottage-ish, whimsical, bijou. For years I felt literally diminished, taking up less space in the world in a house occupying less than an eighth of the square footage of our former home's gaudy splendour.

On the other hand, our new residence wasn't just any old house. I bought an absurdly eccentric Victorian Gothic folly built in 1850. Called 'The Dragon House' in Lord Longford's anthology of architectural photography, because a strange dome decorated with art-deco stained-glass lilies protrudes from the front, like a dragon's beady eyes, it looks like Northanger Abbey's and the Addams Family house's much prettier baby.

Career? What career?

You've read about celebrities' careers plunging into the khazi. It happened to Cilla. It happened to Shane Richie. One minute they're hosting the National Lottery, tucking into a full Irish at Eamonn and Ruth's (pre-divorce) and puking on the pavement outside the Chiltern Firehouse, hair held back from hurtling vomit by a lissom *Love Island*er. The next they've disappeared without trace. Even beloved national treasures don't escape the slide to oblivion. Would you believe the late buoyant Dame Barbara Windsor fought through fallow years before she bounced back into Albert Square?

In 2002 my work dried to a brittle crust and crumbled away. My *Daily Express* column kept the wolf from the door, but TV and radio were there none. I felt like Clarence, the ghost in *It's a Wonderful Life*. He's allowed back to earth to show Jimmy Stewart a few home truths but can't join in the fleshly frolics because he's a mere spectre. I'd been banished from the Magic Kingdom. I contemplated requalifying as an English teacher. I considered taking in washing. I thought I'd be rather better at the Oldest Profession and wondered what the going rate in cash was for hand-jobs. I plotted sending the children up other people's chimneys or chopping up the furniture for firewood.

Just think about the once famous faces who simply evaporated into the ether. Once they sprawled all over your screen as if they had a God-given right to dispense opinions on gender-neutral toilets. Now they languish somewhere called 'real life' and no one gives a fig what they think about proportional representation. It doesn't happen to everyone. Richard Madeley is one of the great survivors. Yet a spell of obscurity, often infinite – imagine stretching till all eternity in hideous anonymity – is a frequent postscript of fame. By June 2002 I was heading for 'Hello. Were you once Vanessa Feltz?' I wasn't over the moon to be a has-been at forty, but I had been expecting it. Remember, the Good Doctor predicted I'd last no more than seven years in the spotlight.

62

Back at the BBC

BBC Radio London

In December 2002 I heard from BBC Radio London's boss. GLR morphed into BBC London 94.9, eventually becoming BBC Radio London. 'Someone's off on holiday. Do you fancy a couple of days' weekend cover?'

Did I fancy? I'd have paid for the privilege. Broadcasting in my old *Jewish London* studio after an eight-year gap? I couldn't wait. Ill-wishers, many of them female journalists who'd taken my ascent as a personal insult – 'What? Giving that fat thing a frigging TV show? It should have been me. I write better and I haven't eaten since 1973!' – were keen to point out the humiliating nature of my dizzying descent. 'Just eighteen months ago, Vanessa Feltz was presenting her own five-days-a-week, network TV talk show,' they crowed. 'Now she's sunk to the downmarket drudgery of local radio. What an embarrassment!'

They didn't get it. They didn't get me. What I thoroughly enjoy, and always have, is the job. When you're on air you're on air. The feeling is the same whether you're hosting *Newsnight* or playing dedications on hospital radio. You are speaking directly to your listener. NB You always think of the listener as one person. You don't want to do what prime minister William Gladstone did to Queen Victoria, who spluttered: 'He addresses me as though I were a public meeting.'

You might be paid more for presenting or, in this case, less. You might have spent an hour in Hair and Make-up, wearing vintage Versace sourced by your personal stylist or travelled on the Jubilee Line in trainers and tracksuit. Once the green light is on, you are on. Adrenaline pumps. You are focused, concentrating, living your best life. Broadcasting is colossal fun. I love collaring MPs and ministers, lobbing listeners' questions at them and doing my best to stop them wriggling out of answering. I've lived in London all my life. I am a Radio London listener, and it didn't strike me as the least bit infra dig to present a show on one of my favourite stations. I'd only been booked for a couple of days' deputizing, but I prayed it might lead to more. Who knew it would take war breaking out in Iraq on Thursday, 20 March 2003?

Danny Baker

Danny Baker, then presenter of the station's *Breakfast Show*, was and will always be a law unto himself. Come hell, high water or carnage in Iraq, Danny would present a show about the weird stuff you keep in your kitchen drawer and the sweets of his youth – Aztec bar, anyone? No editor bleating about covering international conflict could budge Baker. He flatly refused to mention current affairs, toppling statue of Saddam Hussein or no. There must have been a BBC edict compelling all stations to go hard on Iraq. Danny was not for turning. Friday, 21 March 2003, I was called to step in. It was the beginning of a love affair with my Lovely Listener that is still passionate to this very day.

In April I was given my own BBC London weekday show from three to five p.m. It was frivolous. Is a pre-tied bow tie vulgar?

Chunky or smooth marmalade? What should adorn the fourth plinth in Trafalgar Square? I was content with trifles and fripperies. I didn't hanker for hard-hitting political debate in the slightest. I didn't know what OPEC stood for and I didn't care. My parents thought my lack of political interest branded me profoundly superficial. They were right. I am.

The boss kept asking me to cover *The Breakfast Show*. Danny had moved slot. His successor, Jon Gaunt, was going for the jugular in a shouty manner. I politely refused. I know *Breakfast* is the flagship show of any radio station, but I didn't fancy presenting it. I was tickled to death with my afternoon show, alighting gracefully on pleasing topics. My 'virtual advertising agency' propelled Londoners to creative heights. One of my listeners concocted the perfect greengrocery slogan: 'Bananas – wrapped by God'.

Gordon Ramsay

I had a right laugh interviewing showbiz types. Gordon Ramsay, a thoroughly good egg, burst into the studio oozing bonhomie. As soon as we went live, he let fly a surprising volley of F-words. Ashen-faced, the producer hurtled in waving a bit of paper inscribed 'G MUST APOLOGIZE. IMMEDIATELY. OR WE'LL ALL GET FIRED. URGENT.'

I stopped Gordon mid-flow. 'I know you didn't mean to offend, but some people greatly dislike foul language. Would you mind apologizing and please don't do it again.' Gordon couldn't have been sorrier or more obliging. He apologized unreservedly. He'd learned his lesson. I asked him something vanilla about vanilla. Somehow he worked in the word 'motherfucker'. He was contrite but, to our sorrow, we had no choice but to show him the door. He exited with abundant charm. Never has a nicer man been banned from any show.

Michael Gove

I was given a Saturday-morning slot. I was now on air six days a week. Frankly, I needed the money and the girls understood. On Saturdays I featured a notional 'newspaper', the 'London Mouf'. Listeners were my local correspondents, ringing in with spontaneous weather and travel news plus perky 'opinion pieces'. My virtual co-editor was a dazzling journalist by the name of Michael Gove. He was larky, sparky and a thoroughly good sport, who somehow never got round to mentioning his personal political ambitions. When he was reincarnated in 2005 as a serious politico and went on to be a Tory grandee responsible for inflicting a grammar curriculum on Britain's youth and knifing Boris Johnson between the shoulder blades, I wasn't even slightly surprised. He should not have confined himself to churning out leaders for *The Times*. Twenty years later I wrote about our 'on-air' japes in the *Express*. Michael was in touch in seconds. Elevated to *Spitting Image* status, he'd forgotten he'd ever been a puerile prankster and was as happy as a clam to be reminded. We've chomped cheerily through a few dinners since. We still never utter a political syllable.

Russell Brand

Russell Brand asked to co-host my Saturday-morning BBC Radio London show. It was an outside broadcast in a shopping centre in Croydon. Why? We were pals and he felt like it. I met him in 2004 when I was a panellist and he the new presenter on *Big Brother's Big Mouth*. I was enthralled. He was a mercurial maverick playing fast and loose with faux-Shakespearean language, savouring stream-of-consciousness soliloquies laced with double entendres. He said he was trialling his pre-Edinburgh stand-up routine in a pub in Camden that night. Would I come?

I went and was gripped by the aggressively gritty monologue carving welts out of his drug addiction: introducing his dealer to Kylie Minogue, giving a tramp a blow job for blow and spitting as a routine sex act. I didn't laugh much but I was mesmerized by his storytelling. We became friends, friendly enough for him to message me from the bath and borrow my black eyeliner at *Big Brother* founder Peter Bazalgette's party. I liked Russell tremendously. I thought he was kind and wildly, hectically entertaining. I wrote about his tortured genius in my columns. This predated his boasting about shagging *Fawlty Towers*' Manuel actor Andrew Sachs's granddaughter on Sachs's home answerphone. He was egged on by Jonathan Ross and the unedifying episode broadcast on Radio 2. He hadn't yet swirled off to Hollywood, starred in the remake of *Arthur* and married Katy Perry on elephant-back. Yet had you asked me if Russell, lowly presenter of a *Big Brother* spin-off show at Elstree Studios, might do any of those unimaginably epic things, I'd have said 'Yes' without hesitation.

We stopped being friends when, in front of a baying audience on his show *1 Leicester Square*, Russell said he'd like to have sex with both my daughters. The audience roared approval. His tone was predatory. He doubled down. I couldn't pretend to find it funny, even for the cameras. It turned out he'd seen my younger daughter on her way home from school and asked her for a drink in a nearby pub. She said, 'No, thank you, but I know you. I'm Vanessa Feltz's daughter.' She was fifteen. Russell was fascinating, ricocheting with charisma but, sadly, no longer friend material. I doubt he noticed.

BBC Radio London

My boss wasn't prepared to leave me contentedly blathering away about hemlines and hemlock in the teatime slot. He insisted on shunting me to mid-mornings, nine to twelve. That meant addressing the day's hard news stories, with the Lovely Listener to guide me through the choppiest waters. Politicians, professors and pundits were booked as expert guests, but phone-ins were our bread, butter and jam.

Fielding 'pro' and 'anti' calls is dreary broadcasting. 'I loathe ULEZ!', 'I love ULEZ!' Boring! What I want to know is why the Lovely Listener thinks that way. What has happened in their lives to make them hold such an intractable view? I want flesh on the bones of their call, detail, information, personal stories, not deadly dull 'Yes!' and 'No!'

Getting the question right is key. The art is to come up with an angle the listener can't ignore. What will Brexit bring you? Why do you think the NHS is institutionally misogynistic? Why do you believe smacking your children is your right? Why are you childless by choice? Is everyone a little bit racist? Have you been stopped and searched and, if so, do you believe stop-and-search works? Are you leaving London as part of 'white flight'? Will you be alone this Christmas? Do you love the capital's green parakeets or see them as foreign interlopers? Is your nest empty? Is there such a thing as 'conscious uncoupling'?

Phone-ins

I began to piece together pictures of my Lovely Listeners from the fragments of experience they shared. I learned 'Kim from Lewisham' is a refugee from Kosovo battling PTSD; 'Maureen from Plaistow', a wise, humorous, salt-of-the-earth Cockney great-grandma, still mourning the loss of her adored husband Fred; 'Nigel from Temple Fortune', a hard-bitten press photographer with a house in Las Vegas; 'Shane from Edinburgh', an analytical LSE graduate with a wry take on Tories. To this day I use the late

'Joan from Wallington's' Christmas turkey 'build a foil pagoda and give the bird a Turkish bath' method.

Every presenter attracts a coterie of 'regulars'. I loved mine dearly but used them on the show sparingly. You never want first-time callers to be discouraged from ringing because they're put off by what feels like an exclusive gang. The perfect programme features new voices, first-time callers with unheard stories, peppered with a few regulars for a dash of familiarity.

Perfection depends partly on the news agenda. The MPs' expenses scandal made scrumptious fodder. Every day brought freshly infuriating detail. How we loved to hate the Honourable Members who had snapped up duck houses, moat cleaners and Venetian escritoires. Momentum built beautifully. Calls stacked up. Screens flashed emerald – a phone-in presenter's paradise.

Slow news days

Of course, hosting a daily show means pretending not to plod through the quagmire of dire news days. Nothing of note is happening. You managed to scrabble together a show of sorts yesterday, but pickings are slim, and you won't be able to eke out the same thin subject today. You can't open the show: 'Morning. Rotten news day. Nothing you'd care less about is going on, so goodbye. Let's all watch daytime telly and I'll see you tomorrow.' It's your job to conjure something worth listening to.

Sometimes, if you get the question right, no news makes riveting listening. Still vivid in my memory is the call from a pensioner who survived a back-street abortion on a filthy kitchen table. She said it was the first time she'd ever spoken to a living soul about it. I'll never forget the woman sexually abused by her neighbour as a child. She'd tried and failed to explain to her parents why she didn't want to go next door and help the 'nice' man feed his rabbits: she was too young to have the vocabulary to explain what the 'nice' man did to her when she got there. Her parents, not wanting to be rude to the friendly neighbour, forced her repeatedly into his clutches.

I spoke to callers who fell in love, literally, at first sight and were still passionately enamoured decades later, callers who took two buses and a train to their dear departed's grave every day, ate a picnic lunch at the graveside and travelled home again, callers who had been on a council house waiting list for twenty-five years. I spoke to a Croydon teenager terrified to leave her flat at the beginning of the school summer holidays because postcode gang warfare made her a target if she crossed into 'rival territory', and callers who said there were no gangs in delightful London, she must be imagining them. Thank you, Lovely Listeners, for your insights, confidences and revelations. I am immensely grateful to every single one of you.

7/7

On 6 July 2005 (most of) London rejoiced at our unexpectedly successful Olympic bid. David Beckham danced a jig. Our phone lines went berserk. I knew I was in for seven years of 'Bring it on! I can't wait!' versus 'What a crock of shite! We can't afford it!' calls, and that was before Mayor Ken Livingstone said the Olympics wouldn't cost each household more than the price of a Walnut Whip. My briefing call from my producer Esther Stanhope on 7 July was interrupted by shouting and what sounded like a stampede. Esther disembarked the replacement bus service to King's Cross. She'd been told there was a power surge. She saw people running out of the station with sooty faces, others sitting on the pavement crying. She recalls my saying: 'Darling, you're pregnant. Go home now!'

She headed up Pentonville Road and heard an almighty bang. It was the bus bomb at Tavistock Square. She had no idea what had happened.

In the thick of the London rush hour, at 8.49 a.m., three bombs exploded: one on the Circle Line between Liverpool Street and Aldgate, one on the Circle Line between Edgware Road and Paddington, one on the Piccadilly Line southbound from King's Cross to Russell Square. At 9.19 a.m. a code amber alert was declared by London Underground. An hour later another bomb exploded on the top deck of a number 30 double-decker bus in Tavistock Square. Apart from the bombers, fifty-two UK residents of eighteen different nationalities were killed and more than seven hundred people injured in the attacks.

At nine a.m. on 7 July 2005 we didn't have that information. All we had were ghostly shots of shocked commuters emerging bleeding, dazed and wounded. Ambulance sirens filled the air. I live in NW8 two miles away. We could hear them. We didn't know if these would be the first of hundreds of detonated bombs all over Britain. We had no idea of the death toll. What we did know was that we didn't know. People who had waved their children, partners, friends off to work less than an hour earlier didn't know if they were alive, hurt or dead. Telephone lines went down. Communication stopped. The transport network stopped. For hours people were trapped in a nightmare of not knowing. Radio and television were the only sources of information.

I drove to BBC London to be on stand-by. Disaster demands bespoke broadcasting. You must be reassuring, but not falsely so. You must not ramp up whatever is happening and make it worse. The presenter's job is to relay information clearly and calmly – despite the inevitable mayhem in the studio – repeat facts, deliver breaking news and make certain not to deal in conjecture or misinformation. It's a huge responsibility.

Unable to get home any other way, commuters began walking the miles back to their houses. Families with no phone contact could only wait in anguish for the sound of a key in the lock. We broadcast Mayor Livingstone's speech: 'This was not a terrorist attack against the mighty and powerful. It was not aimed at presidents or prime ministers. It was aimed at ordinary working-class Londoners, black and white, Muslim and Christian, Hindu and Jew, young and old. It was an indiscriminate attempt to slaughter, irrespective of any considerations for age, for class, for religion or whatever. This isn't an ideology. It isn't even a perverted faith. It is just an indiscriminate attempt at mass murder.'

Emerging from the studio after four hours on air, I was surprised to see people smiling in the sunshine, licking ice creams and sipping cocktails. In the newsroom we were absorbed in the unfurling tragedy. In the street, life ticked on as usual. Broadcasting often feels oddly unbalanced. Inside the building, focus is funnelled in one direction. Outside most people function blissfully unaware.

Ken and Boris – but not Sadiq

London's first mayor, Ken Livingstone, took questions live on my show every four weeks for both terms of his mayoralty. Always a little testy, living on his last nerve, querulous and complaining of back or neck pain, Ken treated callers crustily. No one could have called him a schmoozer. When a disabled listener complained of difficulty using the tube, Ken snapped crisply: 'When the Underground network was designed if you weren't able-bodied you stayed indoors.'

Ken was so convincingly omniscient I couldn't tell if he was genuinely master of all knowledge or just gamely making things up. 'There's a broken paving stone outside my house, number five Cherry Tree Lane,' a caller would say indignantly.

'Ah, yes. I know that rogue paving stone, right next to the horse chestnut,' Ken would say sagely. I'd stare into his eyes, trying to work out if he really was personally acquainted with the slab. He was inscrutable.

Ken – by now an MP, no longer mayor – hurtled from political grace on my show. On 27 April 2016 Bradford West Labour MP Naz Shah made a 'wholehearted apology' to the House of Commons over Facebook posts appearing to suggest Israelis should be deported to the US. 'I accept and understand that the words I used caused upset and hurt to the Jewish community and I deeply regret that. Anti-Semitism is racism, full stop.'

Ken Livingstone was booked on my programme, simply to echo her apology, or so the team and I thought, and reiterate Labour's pledge to stamp out racism. Instead, with no prompting from me, he said: 'When Hitler won his election in 1932, his policy then was that Jews should be moved to Israel. He was supporting Zionism before he went mad and ended up killing six million Jews.'

I kept silent. Nature abhors a vacuum, so I shut up and allowed him to commit political suicide. He was confronted by Labour's John Mann outside the BBC immediately afterwards and accused of being a Nazi apologist. By that afternoon Jeremy Corbyn suspended him from the party. Ken refused to apologize. On 4 May 2018 he admitted calling Hitler a Zionist had probably cost Labour the chance to snatch Barnet Council.

Boris Johnson

Boris continued Ken's tradition of dropping in once a month to answer questions on my show. He was a huge hit with the team. If the milk ran out, they'd put in a call to Bojo who'd stop off on his bike and pick up a carton on the way in. Helmet askew, bicycle clips akimbo, shirt untucked, grinning broadly, Boris would bowl into the building, often with Guto Harri, his director of communications, chugging along in his wake. Adore or abhor his politics, from a presenter's perspective Boris was an unbeatable guest. His rhetorical flourishes inevitably earned us front-page headlines in the *Evening Standard*. He'd sprinkle his aphorisms with arcane vocabulary, 'riparian' or 'catechism', for added spice.

In October 2010 he responded to a government cap on housing benefits: 'Never shall we have Kosovo-style social cleansing of London. On my watch you are not going to see thousands of families evicted from the place where they have been living and put down their roots.' Hey presto, outrage from all and sundry, and accusations of – guess what – 'inflammatory language'! Boris was – and is – a gift to any presenter. He is incapable of being boring. Did I wonder if his verbal volleying was sport and if he cared more for the performance than the problems he was meant to be addressing? Sometimes. Mostly, watching him was a masterclass in Kipling's 'common touch'.

Sadiq Khan

What can I write about Sadiq Khan? I presented the BBC's flagship *Breakfast Show* for London. He was (and still is) Mayor of London. He refused to follow in either of his predecessors' footsteps and join me to answer Lovely Listeners' questions. He never once accepted my cordial invitation. Was he, as some have suggested, too scared to face me? I don't know. Is the man a lily-livered milquetoast? I couldn't say. All I know is I bumped into him on the tube in 2024 and thought he might have looked a trifle perturbed. He might not have. Maybe he always looks like that.

For God's sake, Professor, rev it up a bit!

Phone-ins usually have three subjects per show. If one flies, you can always ditch the others. One morning we were tackling the possible outbreak of swine flu, the deteriorating quality of language used in children's television programmes and asking if anyone other than Dame Joanna Lumley fancied the idea of Boris's Garden Bridge?

My producer typed: 'Professor XX', on my screen.

I went in hard: 'Morning, Prof. Good to have you on board. Why is the vocabulary used in children's drama so limited? Why have programme-makers decided to constrict children's lexicographical horizons? Is this dumbing down?'

The professor was non-committal: 'I – I – I – I'm not sure, really. Could it perhaps have something to do with the internet?'

I was in no mood for a wishy-washy expert. 'Come on, Professor,' I said impatiently, 'for Heaven's sake, rev it up a bit. Climb off the fence. We want to know. Who is letting our children down and why?'

Ashen-faced, my producer Jonathan Lampon ran into the studio, brandishing a piece of paper. 'STOP! He's the swine-flu expert!'

I grovelled abjectly. The Lovely Listeners never let me forget it. I will gloss over the show in which I somehow merged a delivery of the new harvest from the Asparagus Growers Association with an interview with the Asperger Autism Association. It's easily done. If you don't believe me, try discussing asparagus and Asperger's without making an arse of yourself.

63

One Hit Wonder

IMAGINE GOING OUT WITH A One Hit Wonder. Stretch yourself further. Imagine going out with a One Hit Wonder who didn't write the hit, wasn't there when it was written and was paid a one-off session-singer fee to sing it. Every time the tune blares from the radio, you can't even think: Fabulous. That's another fourpence in the family coffers. Your partner doesn't get a penny from the song that didn't quite make his name. It was created by a band, let's call them 'Big and Little', and he is neither. Big and Little are sitting pretty in houses purchased with the revenue from their stonking tune. They are sitting pretty, but they're not very pretty. That's why your chap had to sing the song and front it up in the video and on *Top of the Pops*.

He's a pretty face. Whether he's 'just' a pretty face, you don't yet know. You've just got together. He seems busy. He's making a one-off appearance as a contestant on a Channel 5 talent show. He doesn't quite portray it that way. The rehearsing and costume fittings give the impression the gig's a full-time job. When the true nature of the work emerges – he performed a few minutes of elementary magic and didn't win – he seems not to do much else. You don't like to press the point and ask what he does for a living. It would feel pushy. Anyway, there's that terrific song on the radio again, and he did say, 'I live off my residuals,' so you are prepared to punch the air every time you hear it, smile broadly when people say: 'Oy, mate! That's your song! Oggy! Oggy! Oggy!' It takes a while for you to realize he was the (pretty) face and synthesized voice, but – unless he's in a club in Essex, or at Butlins being paid to perform the track, which doesn't happen very often – it doesn't earn him anything at all.

The song was a hit in 1999. It got to number two in the UK charts, unfairly – as he told me many times – banished from the number-one slot by Martine McCutcheon's foray into pop music. It stayed at number two for weeks. Everyone loved it. OHW bobbed down to Brighton and, for a few quid, laid down the vocal. When the song took off months later, they gave him a call, and whenever it needed singing, he popped out and performed it. He warbled with welly, in a gold jacket – which he later described as 'iconic' – all over Europe, on umpteen TV shows, on *Top of the Pops* and in every nightclub worth its salt from Uzbekistan to Ibiza. He was twenty-seven. He was on the road with two female backing dancers, living it large, singing the song everyone wanted to hear. He was number-one friend to the stars, hanging out in London's private members' clubs with rockers and boy bands, women on tap, swigging free Dom Pérignon straight from the bottle. What did it matter if no one knew his name? He was killing it.

There was just one problem. Big and Little were never able to replicate the track's success. OHW sang a couple more. One reached number seven, the other number eleven. Then Big and Little replaced him. When we collided at the *OK!* magazine Christmas party chocolate fountain in 2005, demand was already slip-sliding away. We didn't become an item till September 2006. We were friends without benefits. I wasn't sure if he fancied me. He looked so young I worried I was old enough to be his mother. I didn't waste time fancying him. It turned out I was ten years older, not technically maternal territory.

If I thought of him, which I didn't much, I thought he probably thought of me – if he ever thought of me – as an entertaining storyteller. There were hints of flirtation, but nothing convincing. I was relaxed with him. He was good-natured. We ate occasional convivial lunches, the odd congenial dinner. I had no intentions. I didn't think he had either.

My six-year relationship with the Angel of the North was grinding to a halt. He told me he was a nomad. I didn't listen. *Note to self*: if a man looks you in the eye and tells you the truth, do yourself a favour and pay attention. Do not, repeat not, think: Ah, yes, you might be a wanderer who likes to flit from woman to woman and set up tent under the stars, but I will change all that. What you are really yearning to do is settle down with me.

I didn't consider OHW a potential new incumbent. I just liked his jolly attitude. He didn't seem to be wrestling interior angst. I'd had enough of damaged blokes desperately grappling their inner demons. Refreshingly, vodka and slimline tonic in hand, OHW didn't appear to have any.

On day two of a holiday in Italy with Angel of the North in August 2006, I went right off the home-baked five-star breakfast buffet. I couldn't face it. I couldn't face my phone. Looking at it made me feel sick. Looking at the swimming pool made me feel sick. Looking at A of the N . . . You get the idea. My head swam. I was feverish and breathless. I was so ill I was able to sit through two operas in Verona. Healthy, I'd have committed hara-kiri with just one Cornetto. A of the N wasn't exactly Flo Nightingale.

I extracted antibiotics from a local doctor but could barely drag myself on to the plane home. My GP admitted me to hospital. Water made me spew. The paint colour on the walls made me vomit. I was on a drip, sick of being sick. OHW sent me a selfie. He wasn't to know looking at the phone made me heave. His whole family were praying for me. I was grateful for his family's intervention with the Almighty. They were born-again Pentecostal Christians. The consultants expressed concern. It wasn't a patch on mine.

I was *platz*ing (fretting) to the max, not because I'd spent a fortnight gazing at the lavatory bowl but because ITV had just offered me a new TV show: *Vanessa's Real Lives*, a series of twenty programmes involving 'real' people (not pretend ones), was due to start recording in two weeks. Here's a TV life lesson: you can host a telly show with broken limbs, a fractured heart, two days after an operation with a six-week recovery period, with giant fibroids causing copious vaginal bleeding – I have done all four. Viewers didn't suspect a thing. You cannot, however, host a TV show while puking, with a raised rash make-up cannot hide, or with a non-firing voice box. I have tried all three with mixed results. I was pleased OHW was praying for my speedy recovery, and when I staggered from the hospital – the diagnosis was 'unknown virus' – and managed to stay just about upright through the twenty shows, I held him partially responsible.

The platonic phase, with a soupçon of flirtation, ended when OHW kicked matters up a notch. He invited me to hear him sing – no prizes for guessing the repertoire but it was refreshingly new to me then – at the Café de Paris, just off Leicester Square, call time midnight. I was crawling down Oxford Street, stuck in a night-time traffic jam, couldn't find a parking space and was ready to turn around – geddit? – and go home. OHW phoned, his usual semi-somnolent tone replaced with anxiety. 'Where are you? I can't see you.'

'Parking's a nightmare. I think I'll go home. I'll see you some other time.'

'No! Please come. I really want you to be here.' He was such an affable fellow.

I parked on a double yellow. We went from chums to snoggers that night, culminating in a little light grappling in my car. At four a.m. I threw in the towel and drove home. He was horrified. He was, he informed me, 'a nightclubs man'. Days didn't start till mid-afternoon and nights didn't end till breakfast with a couple of vodka chasers. I'm an early-to-bed-early-to-rise woman. Don't tell me. I know. I should have seen a raft of red flags waving right there.

I didn't spot them because OHW uttered the four words in the world I most wanted to hear. 'Tomorrow night at seven?'

What? I hear you cry. You read English literature at Cambridge. You quote reams of Yeats and Andrew Marvell. You recited bits of Harold Pinter's *The Homecoming* to actual Harold actual Pinter. Pinter didn't pause. He praised. Your father sprinkled 'pulchritudinous', 'hyperbole', 'lachrymose', 'chiaroscuro' and 'crepuscular' all over his warehouse of brushed nylon pyjamas. Surely, Vanessa, a paltry perfunctory pitifully pedestrian 'Tomorrow night at seven?' wouldn't send you into paroxysms of ecstasy?

You ought to be right, Gentle Reader. I should have required a sonnet, a villanelle at the very least. I'm not justifying, just explaining. Take one decamping husband. Add to the mixture one dead mother, one disappearing father, one nomadic boyfriend, and somehow the words 'Tomorrow night at seven?' are balm to the soul. Each time OHW said it, and he said it every time, I unfurled a little more. The phrase meant I had an arrangement to see him cemented in place without the need to spend all day worrying, checking my phone or trying to stop myself ringing him. I was signed up in advance to an evening of kissing, cuddling and light-hearted tomfoolery. I badly needed tomfoolery. I'm not sure exactly what it means and can't stomach a Google definition, but tomfoolery was just the ticket. OHW had a comprehensive knowledge of nineties' boy bands and not much interest in anything else, except, he said, the Bible. He was universally acknowledged to be exceedingly handsome, ditto my dad. Reproduced for your delectation a note from my editor: 'This seems an odd moment to refer to your dad in the context of talking about your boyfriend.' My reply: 'Do you think so? Have you never heard of an Electra complex?' Life is odd. I am odd. I have never pretended to be otherwise.

Our coupledom came to light. *Loose Women* did a phone-in about age-gap relationships: 'She's ten years his senior? Can it last?' The papers preferred the race angle: 'She's white. He's black. Can it last?' Every hack with a laptop had a bash at deconstructing the 'unlikely union'. Their conclusion, naturally: 'He's in it for the dosh. She's in it for the sex.' It was obvious to all. Except, at the time, I thought they were wrong on both counts. There wasn't much cash left after the divorce and I was looking for a proper partner, not a gigolo on whom to lavish pearl tie-pins.

When the relationship imploded sixteen years later, OHW told any outlet willing to pay him I considered him 'eye candy' and never intended to marry him. Having survived this book thus far, Gentle Reader, you know only too well that couldn't have been further from the truth. I wanted to be married. I was bred to be married. Six years after a crucifying divorce, however, I couldn't afford to give away half my house to a fellow who sang a song seven years earlier and considered himself too huge a star to do a day job. I wanted a partner who would work and pay his way.

When OHW proposed romantically with fifty red roses spread on the bed of Bono's Clarence Hotel in Dublin, I accepted. Days later, he scheduled a meeting with *OK!* magazine to discuss coverage of the wedding. They mentioned a fee. He eagerly shared his plans for festivities of Sultan of Brunei-style lavishness. That's not how it usually works. Most celeb couples who 'sell' access to their weddings to a magazine throw a modest wedding and pocket most of the dosh. 'Are you sure?' I queried. 'Blenheim Palace? Arrival in a helicopter? Five hundred guests? Don't you want a small wedding and to save the rest of the money to live on?' He didn't.

Later that day I told him I couldn't marry a man with no job. He asked me to stay engaged. I said fine, and I'm always open to a trip up the aisle whenever you work out what you want to do and do it. I still had two children to put through university. I was forty-four and not up for subbing an able-bodied six-foot-four bloke ten years my junior, but I still felt guilty in case I'd led him up the garden path. I gave him back the money he'd spent on having a diamond I owned reset for the ring. That was in December 2006. I wore the ring till 15 January 2023.

Was sex what I saw in OHW? Obviously it wasn't an unappealing prospect but, strictly *entre nous*, I can testify my motivation was not purely rumpy-pumpy. I hadn't been short of athletic boudoir action since falling for the Angel of the North. Baby-faced bald chaps with boy-bandy features aren't necessarily my thing. I'm not saying OHW wasn't a babe magnet. I said it just a few paragraphs ago. He just wasn't the horse for my course. I incline much more towards chiselled cheekbones, a resemblance to Moshe Dayan, an indication that the gentleman is leading Socrates' 'examined life' and, if you insist, throw in some dreadlocks for good measure. I know what you're thinking and you're right. More fool me.

I can't reliably tell you what motivated OHW to stick around for sixteen and a half years. Since our break-up he has given a slew of interviews under the shameful soubriquet 'Vanessa Feltz's Cheating Ex', so you've had ample opportunity to form your own ideas. I can speak only for myself. I was warmed by his positivity, friendliness, enthusiasm and the sweet reassurance of 'Tomorrow night at seven?' He seemed to be in love with me. I was pleased to be in love with him. He said: 'Let's just be kind to each other.' I thought that was an excellent plan. He was nice to the girls, then twenty and seventeen, and nice to my friends. I was the same to his. Barely a barmitzvah took place without *that hit* from One Hit Wonder. He was introduced at Cousin Shmuley's wedding as: 'Cousin OHW, the only member of our family who can hold a tune.'

He lightened the mood at Feltz Towers. Three women together can be hormonally hectic and emotionally intense. He'd crack jokes about Dane Bowers or Antony Costa (of boy band Blue). I know what you're thinking. Seriously, Vanessa? Allegra's a senior tax lawyer with a Cambridge degree. Saskia's a teacher and child therapist working in the state system to counter the effects of lockdown on disturbed children. They are bright, switched-on women. Are you really saying you three were basking in bantz about Boyzone? The girls might have been humouring me, but I'd say, 'Yes.' OHW was groaningly funny, but not gratingly so. He delivered on the 'at seven' promise. He was extra eager to swan up red carpets. I've never met a human who loved the attention more. In fairness, though, he also pitched up at funerals and circumcisions without a word of complaint and met Chief Rabbi Ephraim Mirvis so many times, the two hailed each other like old friends.

Sometimes he'd trot off and sing *that hit* at a nightclub. Sometimes he'd fly off to Ibiza or Turkmenistan and sing *that hit* at a nightclub. Sometimes he'd officially be 'working on' some project of no specific nature. Declining demand for *that hit* meant he couldn't make a living simply singing it. I gently asked if he might consider doing something else. How about session singing? Wedding-band singing? A residency in a pub? Teaching singing? Composing jingles? Carrying heavy furniture? Working in a clothes shop? Learning to drive – he never did – and manning a cab? He wasn't keen. 'You want me to stack shelves in Tesco? Oh no I don't!' (Though if it were me, I'd jump at the chance to pay my own way.) 'Forget it! Ain't going to happen.'

Don't think I don't know the idiot here is me. I do. Miss Dy-Na-Mi-Tee would have sent him packing. Fergie of the Black Eyed Peas would have told him to head straight home and take his 'broke ass' with him. Beyoncé would have thrown him out on his ear and J.Lo would have crisply informed him she wasn't his mama. Now I wish I'd channelled a fulmination of feisty females. Then I was loving being loved and aware that if I terminated all the jollity because OHW didn't share my work ethic, I might bitterly regret it.

We jogged along in harmony for some time. Retrospect easily destroys loveliness and replaces it with wretchedness. Life wasn't miserable. There was the comfort of feeling there's someone on the planet you care about who reciprocates. We didn't have a meeting of true minds but who cares about iambic pentameter when there's someone to share Sunday-morning brunch with? While OHW wafted goodwill in my direction, I was happy to punt it straight back. It sounds pathetic written down, but I was grateful to be loved and prepared to back him to the hilt in anything he did – if he ever decided to do anything at all.

Over time his bonhomie ebbed away. First to hit the skids was – you guessed it – 'Tomorrow night at seven?' OHW became the Scarlet Pimpernel: 'They seek him here. They seek him there.' He was elusive and evasive. If I asked to go with him to a party or gig, he'd spout baffling reasons why I shouldn't come. They made no sense. 'You'll only kick off.'

'Kick off? What? I don't drink or get out of control. If I did want to blow my stack, which I don't, it would end up in the *Mail* so I'd never do it. I don't lose my rag. I get quiet if I'm upset. Anyway, why would I want to kick off at a party? We've been together a decade, name one occasion when I've "kicked off".'

'Never.'

'Exactly.'

'You can't come because you're "too big".'

Again, what? He meant too famous to show up at the event he was dressing up and spraying himself with an entire bottle of cologne to attend. 'But I want to come. I want to be with you.'

'Absolutely not. No. Trust me. This one's not for you.'

Gradually, pretty much everything was 'not for you'.

OHW would ask what we were doing that week.

'Nothing on Monday, a screening on Tuesday, nothing on Wednesday, dinner with the Goldbergs on Thursday, girls here for supper on Friday.'

He'd say: 'Right, I'll be there Tuesday and Thursday. The rest of the week you won't see me, yeah.'

It was a far cry from 'Tomorrow night at seven.' I didn't want a walker. I wanted a lover, a friend and someone to help unbung the waste disposal. 'You won't see me, yeah. You wouldn't be comfortable. I couldn't relax if you were there. Stay home. Trust me!'

Trust me! Was it an insignificant phrase, verbal punctuation? Or was it a blaring cacophony of alarm bells? We all know people preface lies with 'to be honest'. We get 'with respect' means 'I have the utmost contempt for you.' Surely I wasn't too thick to twig 'Trust me!' was a sure-fire indicator the bloke was up to no good. Let's put it this way: he thinks he's too famous to work, doesn't bat an eyelid at living rent-free in my house, contributes nothing to household bills, council tax, mortgage, petrol, holidays or entertainment, and is happy for food, alcohol, toilet paper and paracetamol to be paid for by me. He believes he's a 'pop star' and there is a trickle of gigs where screaming women in their forties rekindle their nineties wild child and throw their knickers at him.

In the first flush of what I thought was love he told me: 'To be a star you must have a USP. Women must want to fuck you and men must want to be you. Imagine being some bird from the Wirral. What's the best thing you can do on a Friday? Shag the lead singer in the band. It's something to tell the girls about on Monday. If you can't shag the singer, you shag the drummer. If you can't shag the drummer, you shag the roadie. Women are mad for it. Mad.' This was light years away from the Valerie Feltz Charm School where a warning in neon read: 'Do not sleep around. You'll get a bad reputation, and no decent man wants to marry a girl with a bad reputation.' Bands didn't come to Totteridge. If they had I wouldn't have been allowed to go and see them, let alone fornicate with them. I wouldn't have wanted to – they might have had foreskins. And even if I had wanted to, I wouldn't have dared, in case Auntie Margaret Silver's mother Mrs Cohen saw me doing the walk of shame in the morning and told my parents or, worse, someone else's parents.

Gentle Reader, you may well ask: 'Vanessa, what did you expect? He was ten years younger, a night-clubber, slithering out of the door to places from which he insisted you were barred – though he wanted to accompany you to Buckingham Palace, 10 Downing Street, the BAFTAs and anywhere else that tickled his fancy. Did you really think he wasn't putting it about? Who do you think you are? Halle blinking Berry? What is so special about you that made you imagine this hunk of burning love would be faithful?'

You're right, of course. How could I have been so deluded? He did keep telling me he was a Pentecostal born-again Christian who knew he would meet the Almighty in the next life. He attested to his faith so fervently I'd have been an infidel not to believe him. It's also true that I wanted to believe him. If he wasn't, what was I doing?

64

Gongs, winning and losing, seeking a Vulcanologist

UNLESS YOU'RE ANT OR DEC, who must spend their sizeable salaries entirely on display cabinets, winning awards is bloody brilliant. The bosses love it. The listeners love it. Everyone feels warm and fuzzy. The team get bladdered at the ceremony and cop off with each other.

You take home a trophy. You put it somewhere modest like the downstairs loo. Secretly you visit it every morning and give it a pat. I did manage to hoover up a gratifying batch of gongs, but let's talk about the nights you sally forth, sprayed and contoured by Hair and Make-up – I usually beg: 'Please, Kardashian the fuck out of me!' – and after all the build-up, some other bastard's name is called out. If you lose Geography Teacher of the Year, at least there are no cameras to transmit your shame. If you don't win a National Television Award or a Sony there's a bloke filming your devastation. You try to style it off. Your heart is a smashed avocado. Your Spanx are killing you. All you want is to pull off your false eyelashes and get into bed with a family-sized tiramisu. You hold back the tears, which would leave mascara trails in your foundation. You clap with Olympian enthusiasm, baring your fangs in an ecstatic smirk. Do you care talk pissing SPORT has won the Aria Award for Best Breakfast Show and you have lost? Care? *Au contraire!* You are thrilled. You are more euphoric than you would be if you had won the prize yourself. You are rhapsodic with delight for the pillaging Visigoths at talkSPORT. You simply COULD NOT BE HAPPIER!

You've been robbed. Your instinct is to stomp straight out of the auditorium. You don't. You can't. It's compulsory to linger a few minutes for form's sake. You don't want anyone to think you're flouncing out – though you're itching to flounce. You make sure you're seen sportingly shaking hands with the victor, then cry your guts out in a taxi. You can see I'm familiar with the drill.

Winning Speech Personality at the radio Oscars

When I did win the Sony Gold for Speech Personality of the Year, my pal Matthew Wright talked up the victor with such immoderate praise, I didn't think he could possibly mean me. I was so good at simulating joy in the face of bowel-wringing sorrow. I sat tight, clapping uproariously, until producer Kate Lyons pulled me out of my seat. Chris Evans presided. It was 2009. I looked podgy in red bodycon. I should have been sartorially savvier. The last thing my billowing stomach needed was 'con'.

I was delighted. I rang OHW. I'd won. Would he come and join the after-party in the ballroom of the Grosvenor House Hotel? We're talking about Mr Nightclub, permanently party-ready. Would he jump into a taxi and join the conga? Oh no he wouldn't. He was tired. He was not in the mood. He really didn't feel like it. 'But it's fifteen minutes in a taxi at this time of night. Do come, please. We're having a fabulous time.'

'No. Not tonight. Don't make a thing of it.'

A couple of days later my father rang. I knew he knew I'd won the Sony. My aunt told me she'd told him. I waited for him to mention it. He talked of everything but. I couldn't help myself: 'Dad, did you hear I won the Sony?'

'What Sony? No. I must go. Bye.'

Eyjafjallajökull

When the unpronounceable – I really tried – Icelandic volcano Eyjafjallajökull erupted in April 2010 flights were grounded and travellers were stranded all over the world. 'Quick, we need a volcanologist urgently!' I told the team. During the news break, I checked on our progress. A work-experience lad was doing his level best to book a Vulcanologist. It is possible to be scrupulously polite while explaining you're not keen to discuss Mr Spock, but it isn't easy. Frustrated callers rang to say how awful it was to be stuck at the glorious holiday destination they'd spent the past year saving up to visit. I enlisted philosopher Alain de Botton to explain why a desirable vacation becomes a prison sentence the moment you can't get home.

65

Gastric band

IN 2010 I MET AN old friend. I hardly recognized him. He'd always been a chap of generous proportions. Now he was svelte. His transformation was due to a gastric band. He'd had the operation in Belgium, not to save money as a health tourist but because his surgeon was a bariatric surgery maestro. I was reluctant about going under the knife for sheer vanity. What if I died and left motherless children behind just because I couldn't resist both halves of a Bounty? I asked my surgeon uncle, Clive.

He was emphatic. 'Have the operation. It's not vanity. It's your health. Yo-yo dieting is doing you no good whatsoever.'

There were just two words stopping me boarding that plane to Bruges – 'Fern' and 'Britton'. *This Morning* presenter Fern had had a secret gastric band operation a few months earlier and viewers were in uproar. Fern had implied she was eating healthily and riding a bicycle. The weight dropped off her. I was dieting manically at the time and kept comparing my paltry weight loss to her magical metamorphosis. I was eating only dust and air. I was exercising like a demented dervish twice a day. How could Fern be getting so much thinner so much faster than I was? I wasn't the only dieter wondering. Fern was so skinny she started flogging us Ryvita. You've got to be skin and bone to be Mrs Ryvita. We were in awe.

The Fern thing imploded when some slimeball blew the whistle. Fern, we found out, had slipped off without breathing a word about it to a soul and had a gastric band put in. Her *This Morning* audience was appalled. Dieters everywhere – me included – felt let down. We'd been buying boxes of Ryvita and cycling seismically in the hope of being skinny à la Fern. Only she hadn't done it the kosher way, she'd sneakily slunk off and resorted to surgery. Shame on her!

Fern tried to face down her critics. She defended her right to have any operation she damned well pleased without having to make a clean breast of it to the entire country. She came over arrogant and unrepentant. 'What business is it of yours?' didn't wash with her viewers. They resented her selling them tasteless cardboard crackers – only my opinion, I know some people are Ryvita fanatics – when she'd had her appetite shut down surgically.

Why, you might wonder, did I give a fig about Fern? She mattered because I'd based my career on members of the public being honest with me about their horrors and happinesses and I felt I owed them the same transparency. The last thing I wanted was anyone thinking I'd tried to slope off and have the operation on the *shnide* (sly).

It's difficult to come clean about an operation before you've had it. I was determined to reveal the gastric band's existence the instant it was in situ. I didn't want to wait a couple of weeks to recover in case someone blurted it to the press. To avoid risking 'a Fern Britton' and bringing the nation's wrath down upon my own head, I struck a deal with *OK!* magazine: central stipulation, the article must stress there was no secrecy. I was open and upfront. I confessed the truth. I'd tried and failed to maintain weight loss purely through diet and exercise and was determined to share my bariatric surgical journey. Phew! Engrossed in Helen

Lederer's 2024 autobiography *Not That I'm Bitter*, enjoying her hilarious description of her gastric-band operation I read: 'The only person who had made it work for her was Vanessa Feltz, who had provided her fans with so much detail there was little appetite for any follow-up.'

I chose a gastric band because the surgeon said it was the least invasive bariatric operation. They pop a plastic ring around the stomach, making the aperture smaller. None of your anatomy is chopped out and flung down the waste disposal. If I wasn't happy, I could always have it removed and be back to normal, no harm done. General anaesthetic apart, he said, there was no risk.

No one bothered to explain what life with a gastric band is like. You can eat and swallow any food at all. Five minutes later life becomes a lottery. You might (a) be able to keep down the food you have masticated or (b) need to puke it straight up again. There's no way of knowing which. Bizarrely, though the band is a circle of inanimate plastic, you can never predict what will happen. On Monday you will comfortably be able to eat and keep down a dollop of lasagne. On Wednesday, you try to eat the same lasagne and dash to heave it straight into the lav. Whatever you eat, whenever you eat, even if you eat something you digested perfectly a few hours ago, there are a tense few minutes while the band seems to make up its mind. Will you, won't you, puke? Will you have to charge Sebastian Coe-style for the Ladies? Or will you have no problem clearing your plate?

OHW loathed the band. He took it personally. He was infuriated by the will-she-won't-she? moment. He'd fume: 'Go to the toilet. You know you need to. Go. Just go!' But I might not need to – that's what I was waiting to find out. My gastric band, he complained, destroyed his dining pleasure. He wasn't altogether unreasonable, it was anxiety-inducing but his irritation didn't help an already pressured situation. I was trying not to annoy him, trying not to regurgitate on the table, trying not to eat much. I must have been very trying.

It would have been worthwhile if only the gastric band had lived up to its billing as a magic weight-loss bullet. In fact, the concept and reality are both flawed. With a band you can't keep down any 'healthy' fibrous food that needs chewing. It would take me three hours to eat an apple and I'd bring up every bite. Chicken – no chance. Celery – don't make me laugh. A lamb chop – not in your wildest dreams. Quickly I discovered, irony of ironies, the only foods I could safely eat with a gastric band were the mushy, melty, liquid, fattening variety. Chocolate mousse – vats of the stuff. Mashed potato – knock yourself out. Ice cream – the more the merrier. I hear you loud and clear, Gentle Reader. Vanessa, by all that's holy, why put yourself through an expensive, potentially hazardous operation to slim yourself down, then commit self-sabotage slurping macaroni cheese?

I know it doesn't make sense. I'm just telling you what happened. I was a fat person with a gastric band. I still wanted to eat. Food was still my comfort and consolation. I still craved sugar. Sugary foods slipped down, no problem. Hey presto, gastric band outwitted, figure marginally less voluptuous but still creeping up from a fourteen straight after the op towards a generous sixteen verging on eighteen. Was it worth all the palaver, including a portal inserted under the skin of the stomach into which liquid was periodically pumped to tighten the band, just to look in the mirror and see a size eighteen Feltz grimacing back? What do you think?

66

Radio 2, Early Breakfast

In 2011 I was summoned to the august offices of BBC Radio 2. Boss Lewis Carnie offered me the *Early Breakfast Show* formerly presented by Sarah Kennedy. I'd be on air every weekday from five a.m. until six thirty. What's more, I'd be spending five weeks a year deputizing for Jeremy Vine on the nation's most listened-to news and current affairs show – wiping the floor with BBC Radio 4's acclaimed *Today*.

I was ecstatic. I adore Radio 2. I'd listened since my teens. I remembered Terry Wogan introducing us to Joe Dolce's 'Shaddap You Face' and calling Julio Iglesias 'Joe Church'. I was a fan. I was crazy about Ken Bruce – still am – and wild about the middle-of-the-road, middle-brow, magical music I'd loved all my life. Give me a dash of Bowie, a sprinkling of Wonder, a snippet of Leo Sayer and a few bars of Mott the Hoople and I'm happy. Don't forget the Nolan Sisters, Tina Charles, Luther Vandross and the Bee Gees. Radio 2 was my spiritual home and now I was living in it. Earthly bliss. Joyful! Joyful! Lord, we adore thee.

Two radio shows a day before ten a.m.

Hang on a second, though. What about my BBC Radio London show? Could I forsake my Lovely Listeners and head off without a care to the verdant pastures of Radio 2? Could I ascend to the heights of Europe's most popular radio station and bid farewell to my loyal local listeners? Fat chance!

I hatched a cunning plan. I'd simply do two BBC shows, one national, one local, one straight after the other. I'd present *Early Breakfast* on Radio 2, drive home to Feltz Towers for a restorative forty-five-minute kip, then drive back to W1A 1AA, walk into an entirely different BBC building, named after the late lamented John Peel, and present Radio London's mid-morning show from nine until twelve.

Had anyone ever hosted two BBC shows on two different stations back-to-back? I didn't know. I very much doubted it. I kept expecting some grown-up from management to march me into a meeting to tell me it couldn't be done. Two shows in the same morning? There must be a statute forbidding it.

Chris Evans gives it three weeks

No one did, except my old mucker from *Big Breakfast* days, Chris Evans. 'You'll never do it, Lady V. Four and a half hours on air? I give it three weeks. Four, tops.' Reader, I presented both shows for twelve years. Two days a week at ten a.m., I'd gallop down the BBC back stairs and leap on to a limo-bike to the *This Morning* studios at White City. By the time I chirped up on set with Phil and Holly at ten twenty-five I'd already been broadcasting for four and a half hours. On the bright side, I'd done three paying jobs by ten forty-five. On the downside, I was knackered, jet-lagged and menopausal before Davina reinvented The Change as a sexy, minxy experience. Like all semi-night-shift workers, I was out of sync with the rest of the world.

Four a.m. starts: One Hit Wonder too fast asleep to notice

Obviously, waking up in the small hours isn't an aphrodisiac. Some couples manage to make it work. When Matthew Wright rose at dawn to present *The Wright Stuff* his wife Amelia dutifully went to bed with him at seven thirty each evening so they could share the same time zone. OHW was unwilling to contemplate such sacrifice. He stuck to his night-owl hours like Blu-Tack. I was never there to observe it personally, but my lovely daily Dulce let slip that, in my absence, OHW didn't deign to shake a leg till nine thirty at the earliest. He was in my house, sleeping six hours longer than me, every weekday for twelve years.

67

Darcey Bussell,
'The cha-cha-cha is not your dance'

'Vanessa, will you do a routine as Cher on *Let's Dance for Comic Relief*?'

'No problem.' I took my eye off the ball. I was going to be mother-of-the-bride (not to mention father, grandma and grandpa of the bride) at Allegra's wedding to six-foot-seven-inch dark, handsome French technical wizard Dan at the Plaisterers' Hall just seven days earlier. Planning, paying for and presiding over a wedding takes it out of you. I was responsible for everything and scared of messing up. For months I awoke from nightmares in which I fed the vegetarians lamb tagine, under-tipped the rabbi and ordered a cake instead of a Gallic *croquembouche*. Throwing a gigantic formal *simcha* (celebration) without a husband or mother to consult is daunting. I wilted under the pressure.

I couldn't bear the thought of Allegra moving out and wafting off to Hendon. Dan won her heart in Rome while dressed as a centurion. Don't ask me why. That's just the way he rolls. She was enveloped in a cloud of love. I was feeding hundreds of Dan's Parisian relatives – he has at least two hundred cousins – shoals of freshly cut sushi and dreading the 'Bye, Mum' punchline.

Batmitzvahs are so much better. After 'Auld Lang Syne' your daughter comes home and sleeps in her own bed. Being mother-of-the-bride means spending mountains of money marking the moment the daughter you adore leaves home for ever.

No wonder I wasn't concentrating on the Cher routine. I failed to realize I'd have to wear the iconic fishnet stockings-boots-leather-jacket-nothing-else ensemble topped off with a thigh-length black wig. No one happened to mention I'd be wired up and lowered on to a cannon from a dizzying height, legs suggestively akimbo. 'If I Could Turn Back Time'. My thoughts exactly. I was so nervous my front teeth clattered.

On the night, the girls and, for some reason, Nicole Scherzinger, who couldn't have been more encouraging, were in the studio to watch me bare almost every inch of bulging buttock and keep smiling when an overeager wire operator brought my vulva crashing down on the sodding cannon. It looked as if the audience might vote for me to come back the following week to endure the ritual humiliation all over again. I lost all dignity. 'I'm begging. Please *do not vote for me*. You've seen it. You'll never be able to unsee it. Don't put yourselves through this torture again.'

I thought my dancing days were over. Little did I know my clay-footed homage to Cher would lead to the call of dreams from Britain's all-time favourite programme *Strictly Come Dancing*. I love *Strictly*. I've been a fan ever since Natasha Kaplinsky waltzed off with the very first glitterball in 2004. A minor impediment like two left feet didn't stop me being tickled pink to be sprayed with fake tan, sprinkled with sequins and paired with sharp-witted, sinuous-hipped professional James Jordan. We're still friends. We both knew I'd be one of the first turfed off the dance floor, but James was determined I'd play it straight, not scrabble for laughs Ann Widdecombe-style.

Doing *Strictly* is special. For the audience, which, when you're in the thick of it, feels like every man, woman, child and household pet in the country, *Strictly* is personal. Your dry-cleaner cares passionately about your Argentine tango. Ambulance drivers comment on your heel turns. Checkout ladies correct your 'armography'. Knowing you're not a good dancer and thinking, in the scheme of things, your dancing doesn't matter all that much is not a *Strictly* thing. To fourteen million absorbed aficionados, the left sidestep of your cha-cha-cha matters very much indeed. At times I contemplated commissioning a sandwich board inscribed: 'I didn't go to stage school. I never said I could dance. It's not my fault they asked me.' I didn't because people were universally supportive and generous to a fault.

The *Strictly* curse is unavoidable. Your professional scoops you up in his rippling biceps and whirls you to Paradise. Professionals pilot your every waking moment. They choreograph your dreams. James said, 'Rest your fou-fou on my right thigh.' And that's exactly what I did. It felt delicious. My fou-fou and his taut thigh melded in perfect harmony. For at least forty-eight hours, I fell headlong in love with him. His lissom wife Ola shimmied into our rehearsal room to help me out with a few twiddly bits and I came to my senses, but it's easy to see why contestants' marriages fall apart at the first ravishingly romantic waltz 'in hold'. Among my *Strictly* 2013 alumnae were rugby player Ben Cohen, who later left his wife for his dance partner Kristina Rihanoff and Rachel Riley, who is now married to her partner Pasha Kovalev. I've stayed pals with Rachel, Susanna Reid, Sophie Ellis-Bextor and Abbey Clancy – who went on to win with Aljaž Škorjanec in his first appearance on the show.

I had nightmares about slipping mid-routine, skidding into Sir Bruce Forsyth, toppling him like a skittle and causing him to sustain a fatal injury. I'd wake slathered in sweat, imagining the headline: 'The Bitch Who Butchered Britain's Beloved Brucie'. James and I were rehearsing for eight solid hours a day. I was still presenting *Early Breakfast* on Radio 2, followed by *Breakfast* on Radio London. There was no time to pee or write my newspaper column. Don't tell the *Express* editor, but I had no choice except to learn the arcane art of speaking on the radio and writing my column at the same time. It's like patting your head and rubbing your tummy but with serious repercussions if you screw up.

Strictly week one no one is for the chop. Everyone survives. Cutting me to the quick, judge Darcey Bussell said sniffily: 'Vanessa, the cha-cha-cha is not your dance.' How did she know? She'd never seen me do any other bloody dance. Week two, I was swathed in ice-blue chiffon while James and I managed a lyrical waltz to Whitney Houston's 'Run To You'. Each time James waltzed off over the horizon leaving me alone I wept. I have abandonment issues. I wondered if I'd hold it together on the night. I didn't. The judges were kinder. My waltz was moving and emotional. 'Well done, Vanessa.'

Thank Heaven for golfer Tony Jacklin, who did me the favour of being first celeb kicked off. May his name be for ever blessed. Far be it from me to reignite the controversy over whether I was robbed, and if justice had truly been done designer Julien Macdonald would have gone home second, not me. OK. Let me reignite it a little bit – this is my book after all. Macdonald should have bitten the dust before I did. James Jordan said flinging me to the wolves before him was a travesty of justice. So did my hairdresser. Hundreds of outraged viewers emailed the BBC lamenting my premature departure. I won't tell you if I printed out the emails and still have them hidden in my nightie drawer. I think Julien felt guilty at outlasting me. He kindly sent me a bunch of freesias with a disarming note.

Anyway, eleven years later I'm over it. I've forgotten all about it. Well, if you count seething, aggrieved and thoroughly fucked off as forgetting. Just joking or, to be accurate, so *not* joking. Comedian Rory Bremner said being thrown off *Strictly* is like watching a bus with all your best friends on it, quaffing champagne and eating steak Diane, drive off leaving you behind. He was understating it. Although I didn't have a hope in hell of staying past the first couple of weeks, I was sad when the *Strictly* bus drove off without me. PS I did listen to Sir Bruce – it was his final *Strictly*, God rest his soul – and I did keeeeep dancing!

PS the Amanda Abbington/Giovanni Pernice *Strictly* scandal was news to me. That could be because if you have no earthly chance of winning, your professional isn't fired up enough to bother torturing you, or just because 2013 was the year of love, not war. I found *Strictly* a blend of pressure and pleasure. James and Ola Jordan are still good friends of mine. They munched cheesecake with Rob Rinder, Anneka Rice and Anthea Turner at my last Feltz Towers shindig.

68

All sorts, gastric bypass, Covid, take your pick

Manchester Arena bombing

As I arrived at Radio 2 at 3.30 a.m. on 23 May 2017, news of the Manchester Arena bombing was breaking. What is the right tone on an easy-listening station when twenty-two people, some of them children, have been killed in the deadliest terror attack on British soil since the 2005 London bombings? You can't ignore the news and kick off in normal sprightly style: 'Greetings, Lovely Listener,' chirpy upbeat chatter, 'and here's another classic from Simply Red.' On the other hand, it's not a news station, your job is to uplift and entertain. You can't unilaterally elect to stop playing music and cut to rolling news. I felt the thing to do was express emotion and empathy, acknowledge the sadness and shock, and press on with a record that didn't feel inappropriate. There was no guidance from management. As ever, the top brass were all sleeping soundly at five a.m., oblivious of the unfolding tragedy. Somehow, you must feel your way to hosting a show that is respectful but still true to the character of the station you work for.

Grenfell

The Grenfell Tower fire remains seared in the memory of every journalist who covered the devastation. The interminable inquiry offers no peace, no closure. One of the BBC London execs was playing in a golf tournament on 14 June 2017. People pressed against windows screaming for help before they were burned alive, fire-engine ladders too short to reach the upper floors of the twenty-four-storey tower block, ambulances unable to access the area and the newly appointed Member of Parliament sobbing on air, asking Londoners to get down on their knees and pray, were apparently insufficiently compelling reasons for him to leave the course. We never did find out if he won the match.

Seventy-two people died. Illusions about equality and fairness, if anyone still had them, were cremated that day. Prime Minister Theresa May popped over to view the smouldering ruins and didn't think to get out of her limousine to talk to survivors. Later she claimed she'd been advised not to 'for safety reasons'. Every day I spoke to lost souls, who fled barefoot clutching only their keys and no longer had a lock to put them in. Every day I vowed we would never let the story drop. We would keep it at the forefront of Londoners' consciousness. The inquiry continues.

Gastric band goes wrong

I probably would have soldiered on with the gastric band if, in 2018, I hadn't become ill. I could hardly swallow water and eating anything was pretty much impossible. I also experienced a truly terrifying symptom. I'd be fast asleep and wake up drowning in my own saliva. It flooded my mouth, nose and throat. I couldn't breathe. I felt like Shelley Winters in *The Poseidon Adventure*. I was drowning on dry land. The doctor took it seriously. It's called oesophageal brash. I had a scan. The gastric band had slipped. It was dangerous. It had to be removed. *This Morning* asked me on to discuss it. They booked specialist bariatric surgeon Mr Shaw Somers for added gravitas. After our item, he said kindly: 'You'll have to have the GB removed. And you do know what'll happen then, don't you? You have an eating disorder and you'll put all the weight you've lost straight back on and more. Have you considered a gastric bypass?'

I said: 'Absolutely not.' I'd heard it was a radical and dangerous operation, much too extreme to contemplate. I'd only had the gastric band because it was sold to me as minimally invasive. There was no way I'd dream of a bypass.

Calmly he suggested I reconsider. 'There've been major medical developments. The operation is safe and straightforward. I've performed thousands. Look at it this way. You don't want to be five years older and five stone heavier and come to me asking for a bypass. I should do both at the same time, take the GB out and give you a bypass. Believe me, it'll be the best decision you've ever made.'

Gastric bypass

He was right. This book froths with my idiotic decisions. The bypass isn't one of them. Lord knows how it works. I'm too squeamish to delve into the technicalities. All I know is, you feel full quickly and, once you do, you have no interest in eating another bite. You don't vomit. There's no horrible downside. I'm enormously relieved not to be enormous. I thank Mr Somers for carving out the gastric band, which had embedded itself in my liver – minimally invasive, huh? – and for freeing me from the guilt and self-recrimination of eating too much.

Nothing else to say on weight – hooray!

Since the operation in January 2019 I've been size twelve to fourteen and immensely grateful not to have to discuss my weight with journalists or anybody else ever again. If anyone tries to engage me on the subject, I smile broadly and say: 'Gastric bypass.' There's nothing else to say. Not a word. Hooray!

Covid

Question 1: Do you broadcast in a pandemic? Question 2: How do you broadcast in a pandemic? Question 3: What do you broadcast in a pandemic?

Answers: 1: Yes. You are officially a key worker. 2: You work in a near-empty building keeping yards away from the few colleagues still showing up. Guests, contributors, weather and sports journos are all invisible DTL (down the line). 3: You try to work out which experts are toeing the party line and which are telling what they truly believe is the truth.

When news broke that people in the UK were going down with Covid, one of the execs, a vintage BBC suit with an unerring flair for failing to spot a great news story, asked me how long I intended to focus on 'this illness no one really cares about'? I told him everyone I knew was thinking of almost nothing else and ploughed on. Before lockdown, guest experts, virologists, bacteriologists, consultants in intensive-care units and World Health Organization representatives assured me everyone should mill about as usual.

'Do you mean my listeners should travel on public transport even if they don't have to for work?'

'Absolutely.'

As reports of hospital admissions grew, I pressed harder: 'Are you saying my listeners should go into town and mix in crowds of people this weekend?'

'Yes. Definitely. People should have fun and stick with their original plans.'

I didn't see how this licence to rove could protect the vulnerable from what was clearly a killer disease. So, I changed my line of questioning and made it personal. 'Do you love your mother?'

'Yes, very much.'

'Well, Doctor/Professor/Health Minister, would you be happy for your mum to get on the tube tomorrow and attend a packed concert at the Festival Hall?'

'Yes. I'd be delighted for her.'

Experts continued to declare normal behaviour safe and advisable, until one day, a virologist said: 'Actually, I was due to attend a concert at a jazz club tomorrow and I'm not going.'

EUREKA! 'Why not?' I shouted. 'Why aren't you going?'

'I've been looking forward to this performance for ages. I bought the tickets last year. I'm not going because I'm sixty-five, I have a heart condition and I don't want to be in an enclosed space with hundreds of other people and be infected with Covid.'

Bloody hell! Honesty at last. Now I could refer to him every time an on-message medic spouted the same guff about 'business as usual'. I didn't want to add to the fear factor, but I couldn't have squared it with my conscience if I didn't at least try to unearth the truth.

Lockdown

Lockdown! The concept was anti-British. It might work in Asia, but we were libertarians. We'd never meekly obey an order to stay inside our own four walls. Boris tentatively suggested we stop going to the pub. Sitting next to me on the *This Morning* sofa next day, his dad Stanley said he would be ignoring such nonsense and enjoying a few jars at his local that evening. Next day Boris imposed a national lockdown. We were confined to quarters.

On the sofa, next to OHW, I said quietly, in what I assume was disbelief shared by sixty-five million other Brits: 'Wow! Oh, my God! We're actually in lockdown.'

He was angry. 'Control yourself. Don't make a fuss.' What an inviting beginning to a sentence of twenty-four hours a day stuck in one another's company.

I was lucky. I still went to work every morning. Broadcasters were labelled 'essential workers' so at least I was permitted to leave the house and glimpse my colleagues marooned at safe stations, behind plastic barriers across the barren BBC landscape. Most people paid lip service to the notion of learning something during lockdown, taking up the banjo, baking banana bread, delivering care packages to the elderly couple across the street.

Not OHW. He subsided into a corner of the sofa and wouldn't shift. The balmy weather didn't entice him. My horticultural efforts – I was gardening as if Monty Don himself was about to descend on Feltz Towers and give me marks out of ten for weeding – didn't persuade him to set foot outside and have a look.

OHW didn't do much pre-Covid, and in the pandemic the smattering of gigs dried up. There were no nightclubs in which to 'put your hands up'. He became a sedentary silhouette – up for draining a bottle of wine a night, delivered by Asda, paid for by Yours Truly, but very little else. Would I have detested being locked down alone? Most definitely. I'd have loathed every second. Was being locked down with someone who couldn't bring himself to speak to me, play a card game, listen to music, sit on a neighbouring deckchair in the garden we were lucky to have, any better? You decide. He seemed, as usual, to be living mainly through his phone. I'd finally bought my first smartphone. I needed it to FaceTime my absent grandbabies. Although I began to fall under its spell, I still didn't grasp what kept OHW so permanently absorbed.

The BBC Radio London *Breakfast Show* took on a new function during Covid. Listeners were isolated, frightened and seeking information. We heard from people whose loved ones had been carted off in ambulances, health workers without PPE (personal protective equipment), doubters who didn't know a single person with Covid and didn't believe it existed, business owners who feared their businesses would perish, workers who couldn't afford to isolate, and conspiracy theorists of every persuasion.

The Lovely Listeners and I quickly learned the government's demarcation line between those designated likely to grow very sick or die with Covid – the old or those with underlying health conditions – and those who should be sailing through the pandemic unscathed was, at best, unreliable. We spoke to a formerly healthy forty-two-year-old tree surgeon, who had spent days on a ventilator and very nearly lost his life, a young gymnast who had only just come home from intensive care. We heard from mourners of relatives who 'shouldn't' have been at serious risk of dangerous illness or death but had been killed by the virus.

The diet of doom was unbearable. We decided to try to enliven the gloom with survival stories. The news was death-focused. We would give a platform to the living. 'I Survived Covid' became our favourite feature. We needed sparks of hope in a terrifying terrain. People rang to say they'd been felled by the virus but were recovering. Some were touched so lightly their Covid bout was less gruelling than the common cold. Some had been hovering around death's door for months and were drooping with the exhaustion of long Covid. We were uplifted. They were alive to tell the tale.

Early Breakfast *moves to four a.m.*

Radio 2 controller Helen Thomas was too scared of Covid to sleep. For the first time she joined the 2.7 million listening to *Early Breakfast* on her station. How do I know? She started messaging me mid-programme. Quickly she realized what my darling producer David Manero and I had known all along: my Lovely Listeners were the key workers keeping the country afloat. They were the people the rest of us were banging saucepans for: doctors, nurses, porters, delivery drivers, road and bridge maintenance, police and fire services, retail staff, funeral directors. Once she got the gist of it, she moved my show start time from five to four a.m.

Spring 2020 weather was idyllic. One caller described the blissful forty-five minutes in the morning when her tiny balcony was bathed in sunlight. She was stuck in the shade for twenty-three hours and fifteen minutes each day, but she blessed the Lord daily for her brief gilding in the rays. People took pictures of flowers on their permitted exercise exeats. Listeners shared photographs of straggly daisies and parched dandelions. We all needed glimmers of beauty and hope. We found them wherever we could.

Masks were masked in mystery. Could they protect us? Should we wear them? 'NO!' yelled the experts. 'DO NOT WEAR MASKS! They will embolden you. You will be more likely to contract the disease. And even if they do trap the deadly virus, as you take off your mask you'll smear Covid all over your face and infect yourself.'

'Are you sure?' I kept asking. 'Could they not offer some protection for others? Might they not limit the sheer quantity of virus we inhale?'

'No! No! They offer no tangible protection.' Not one expert would concede masking had any value until the morning a rogue virologist explained that even a porous fabric pricked with holes provided some worthwhile protection. In lieu of purchasing a mask, he suggested listeners moisten a T-shirt and drape it over their mouths and noses in public places. Gradually, the establishment view turned until masks were not merely advisable but essential. We were not permitted to enter shops or public transport without them.

We broadcast the miraculous morning the first British pensioner received her vaccine. Very gradually restrictions lifted. Sports and weather presenters drifted back. Working from home, BBC bosses, of course, largely stayed working from home. We abused the phrase 'the new normal'. It was a barefaced lie. Nothing was normal and we were husks, desiccated and hollow-eyed. We pretended we'd snapped back, like M&S knicker elastic. We hadn't.

One Hit Wonder snoring

I had to leave home for work at three fifteen a.m. Sometimes OHW hadn't returned from nocturnal carousing. Three fifteen and still not home? Were nightclubs even open at that hour? He'd head off for a night on the tiles leaving an empty wine bottle on the kitchen table, downed before a potential nine more drinking hours. He agreed he was a heavy drinker, liked the feeling and could easily stop if he wanted to.

He wasn't a bolshie drunk. He didn't pick fights or upturn tables. He just got, as he put it, 'wavy' and sometimes fell asleep in public. What attracted attention was his dedication to drinking, carrying full glasses to and from rooms, in and out of cabs, once – without success – from a reception at Westminster Abbey into a luncheon at the House of Lords. He didn't look for the waiter in a normal I-wonder-where-the-waiter-is way. He combed the room like a parched impala searching for an oasis.

Funding the drinking was my responsibility. I didn't like it. Seven straight goldfish bowls of Malbec in a London restaurant cost a fortune. I kept coughing up for the sake of what I hoped would be a quiet life.

Friends would say: 'Let's wind up dinner. It's ten thirty. V, what time do you have to be up?'

'In five hours.'

'You poor thing. Come on, let's get the bill and go.'

OHW never said a word. On the way home I'd say: 'Mark Reuben realized I was falling apart. Why didn't you? Don't you care?' Answer came there none. He was spark out as I drove. He never learned to drive. I was the chauffeuse, ferrying a snoring OHW home to help him navigate the stairs before he settled down to a good twelve hours of restorative kip. It wasn't much fun, but at least the snorts and grunts were some sort of company.

I missed them when he started staying without me at the parties and premieres he'd attended as my plus one. At some point I'd cave. I'd be in the sweaty grip of my eighty-seventh hot flush at eleven thirty on a weeknight and admit defeat. 'I need to be up in four hours. I've got to go home.'

'Really, why?'

'I've just told you. I'm on air in five hours.'

'OK. You go. I'm enjoying myself. Bye.'

I'd trail home alone like Macaulay Culkin. Sometimes it made sense for OHW to stay. Take That were playing, and Robbie was back. Of course he wanted to be there. Most of the time, insisting on staying on was downright bizarre. Most people were calling it a night. Frank Sinatra wasn't rising from the dead to sing 'My Way'. OHW wouldn't be missing anything notable if he came home with me. If I suggested we leave together, he'd be exasperated. 'You're just going home to go to bed. What do you want me sitting on my own downstairs for when you're asleep? Why shouldn't I stay here and have some fun, FFS?'

I couldn't fault his logic, but I wanted him to *want* to be with me. I felt like his 'access all areas' lanyard. Comfortably ensconced in the VIP area, red wine in hand, he waved me off so he could get stuck into the real business of Lord knows what without the inconvenience of my presence. When the relationship crashed and burned, people told me his behaviour had been noted.

It's sad when your partner can't wait to see the back of you. It's horribly lonely. I wanted to end the relationship. I was upfront about it. I had a legal letter drawn up and delivered to him – at my own house. It wasn't the first time. The problem was that whenever I was on the brink of packing it in he'd let some droplets of loving tenderness drip in my direction. I'd remember the jokes and affection. He'd revive his sweetness. I'd lose my nerve.

69

A line of bosses. Has someone died?

I win and lose

IN 2017 I WON THE Gold Gillard Award for Best Breakfast Show. Naturally, we played Spandau Ballet's 'Gold'. I glowed all over the airwaves. Rip-roaring *mazeltov*s all round. We were winners. We were merry and bright.

Straight after the two shows on Mondays I wrote my *Daily Express* column. I still do. That morning, I dashed off 1,200 sparky words. I planned to whisk OHW off for a fitting gold winner's 24-carat lunch. Finishing the final sentence, I raised my head from the screen and looked up to find a quartet of BBC suits standing solemnly in a row. They looked funereal. Had they come to tell me somebody had died? My God. Had something happened to one of my children?

A woman had contacted bosses at Radio 2 and Radio London – she was nothing if not thorough – to tell them she'd been having an affair with OHW for the past eighteen months. There was lots of detail. Positions were itemized. She included graphic descriptions of mutual moaning ecstasy. It may sound as if I'm romping through this in a water-off-a-duck's-back kind of way. I'm not. It was a thwacking shock. This was unfolding in the presence of the biggest of my bosses. My producer Gemma Stevenson was doing her best to add a touch of human kindness. Our deputy editor Lorraine Maguire was brought on board to back her up. It was excruciating. I was deeply embarrassed.

When the suits finally pushed off and I'd stopped panting like a winded whippet, I decided to ring Big. He'd DJ'd at the gigs in Thailand and Ibiza where the turbo-charged fornication occurred. I thought it would be pointless. I was sure he'd do that all-boys-together, what goes on at badly paid gigs in second-rate venues on a couple of continents stays, etc., etc. I was wrong. He seemed only too relieved to spill the beans. Yes, OHW had been shagging this bird. Yes, he'd been doing it for a year and a half.

Blimey! I hadn't expected that. I was exactly like any and every woman who has been cheated on. There was nothing at all special about me or my misery. It was same old, same old. Agony.

I rang OHW. He could tell from my voice something was up. On Valentine's Day 2024, fourteen months after we'd broken up, he gave an unkind interview to *MailOnline* in which he claimed I 'stole' the best years of his life. This is what really happened. I drove home and told him the game was up. The broad he'd been knobbing had blabbed to my bosses. Big confirmed her story was true. Would he please pack and go? We'd been together eleven years, and I had no interest in continuing a relationship with a lying cheat. He cried. He raged. He said he'd been blind drunk, and the woman came to his room, entered without knocking and literally prised herself on to his cock. He said he didn't enjoy it. He said he only continued fucking her to keep the Ibiza gig. He said he had to carry on shagging her so she didn't tell Big, who would have been livid. He said he carried on *shtupp*ing her for a year and a half because he was scared she'd tell me. And now she had – WHICH MADE HIM FEEL SO MUCH BETTER!

When he sold his story to the *Mirror* in February 2023, he

described sex with the woman as 'a chore'. What a gentleman! When she sold her story a week later (she would, wouldn't she? No side-piece worth her salt wants her coital skills dismissed as 'a chore'), she rhapsodized about nights of writhing, lust-fuelled passion and pillow talk featuring derogatory comments about me. That wasn't his version. To hear him tell it, she'd practically raped him. As Boris Johnson would say: 'It was a pyramid of piffle.' And while we're on the subject, OHW seemed, to my horrified amazement, exclusively aerated by the fact that, in his aggrieved opinion, Big had 'betrayed him'!

Writing about the cowardly dribble of lies he spouted makes me loathe myself for being such a stupid, suggestible sap. Gentle Reader, if you can stomach another line of this unutterable bullshit, you'll be screaming at the page: 'For God's sake, Vanessa, where's your self-respect? This bloke isn't your children's father. He doesn't pay your mortgage. His conversation is stultifying. He watches Marvel movies. He thinks he's too famous to do a day job and lies in bed like a sultan at your house while you schlep out to work at three fifteen in the morning. You're an agony aunt, FFS. Don't screw this up. This is your out clause. This is the moment you eject him from your home, life and heart, and start to live your life on your own terms. Embrace his exit. Fling him and his thirty-seven pairs of size fourteen trainers to the wolves. Change the locks. Start afresh.'

As usual, Gentle Reader, you are spot on. I should have chucked him out and faced up to singledom on the spot but let me give you the whole woeful picture. He knelt on the kitchen floor at Feltz Towers clutching, I kid you not, a Holy Bible and swearing he'd never stray again. Holding the New Testament to his heaving chest, he vowed eternal love, fidelity, to start taking the rubbish out and possibly, maybe, seeing his way to trying to pay for a few things and, providing I would fund them, of course, taking another course of driving lessons. He was sobbing. He begged to stay. He said I was 'the love of his life' and the drink made him do it. (NB He never uttered a syllable about drinking less.)

I threw him out anyway. He phoned on the hour, every hour, weeping and professing penitence. Ten days later I caved, he was back, and we carried on as before. What on earth made me do it? Anything I write will seem either glib or self-justificatory, but I'll have a bash. I missed the loving, tender times with him. I missed the noise and commotion of his comings and goings. I missed the jokes and, to a lesser extent, the arguments. I missed having someone to do things with when there's nothing at all to do. I had to go to a wedding and didn't want to sit on the – fate worse than death – singles' table. I had booked and paid for a holiday for us both to see the Alhambra Palace and was too much of a nincompoop to go on my own. I loved him. I'd been with him eleven years. I couldn't face another broken relationship. You get the idea.

I had a whole raft of reasons and all of them were bollocks. I should have bitten the bullet. I didn't. I didn't tell anyone else about OHW's affair. I reckoned if we were going to continue as a couple it would only work if my friends and family were kept in the dark. Deep down I must have known they'd be opposed to my carrying on the relationship. Most of them had strong views by now about his behaviour and the information would have tipped them all over the edge. I didn't want to be on my own.

My husband buggered off without so much as a goodbye. There was no kneeling on the floor wailing and begging for forgiveness. I wished there had been. Eighteen years later, OHW was saying he couldn't live without me. Part of me had been longing for someone to say that and, when he did, I lacked the gumption to yell: 'Just get out!'

Taking a partner back after an affair

Anyone who takes a partner back after they've been unfaithful will recognize this bit. They're back. End of. Only it's not the end. You're furious. You're hurt. You're steaming with anger. You want to ask a million questions. You know every answer will be a dagger through your heart and most answers will be lies anyway, but you can't help asking. How many times? Where? When? In what position? Did you have dinner first? Did you spend the whole night in bed together afterwards? Did you think about me? You want details. Each one worms its way into your soul and takes root. You think of him pounding away at her and recoil at the thought of ever having sex with him again. You know if you refuse to have sex with him, the relationship is over, so you're going to have to do it at some point. The prospect is revolting.

You promise yourself not to bring up his infidelity. You bring it up twice before lunch. He manages mild contrition for two or three days. Four days in, he's had enough of atonement. He's comfortable. 'You're not bringing up all that shit again? For real? Pack it in and do us both a favour. I said I'm sorry. What the hell else do you want? Blood? And I just can't believe Big dobbed me in like that. I mean how could he? What kind of bloke blabs about another bloke's dalliance to the bloke's wifey? What about the guy code? What a wanker!' (NB I'm not sure where he came up with the word 'dalliance', but he did, and he wasn't afraid to use it. Six years on, every time he sells his story, he describes his cheating as 'dalliance'.)

I yelled back: 'This isn't about you and Big. This is about you and me. You were unfaithful to me. He wasn't unfaithful to you. The pain is mine, not yours.' What a waste of time. What a squandering of energy. I was a wimp. I should have had the balls to call it quits.

What happened over the next five years is so predictable I grit my teeth to write this. The relationship deteriorated. He was absent more. He was, if that were possible, drinking more. He was glued to his phone like a truculent teenager and his phone seemed to be literally, and I know what that means, glued to his anatomy. It followed him to and from the lavatory. He could not and would not put the thing down for a second. Until Covid I used an £8 Nokia. I'd managed never to go online. I'd never sent or received an email, bought anything from Amazon or watched porn since *Confessions of a Window Cleaner* in 1974. I was still paying my gas bill with a cheque, envelope and stamp. I still do. In a free moment, I'd read, garden or even, can you believe it, think? Resisting a smartphone, still using Directory Enquiries, I'd never fallen down a social-media rabbit hole. OHW had been sucked in so far only the backs of his heels were visible.

OHW's temper was perpetually simmering, sometimes boiling over. What was he so angry about? I had no idea. The scenario was familiar from shows I'd presented. Guests (almost always female) would say: 'I never knew what I'd done to provoke him, Vanessa. Sometimes I'd put the shoes away in the wrong order. Sometimes I'd turned the TV volume too high. I spent my life trying not to upset him, but I never succeeded.' OHW's fury erupted, often in the car, on the motorway when I was driving. He barked, never bit but his roar was so loud, I'd feel as if my brain was knocking against my skull. I was never sure what the catalytic sin would be, but without meaning to, I'd commit it.

70

How can I be sixty?

Sixtieth birthday

ON 21 FEBRUARY 2022 I woke up a sixty-year-old. Old people are always telling us they feel eighteen inside, time fast-forwards when you hit fifty, so 'gather ye rosebuds while ye may'. Naturally, we don't believe them. We think they're antiques jabbering on about boring old shit that has nothing to do with us. Let me issue a warning, Gentle Reader. If you manage not to die, you end up so old you think you must be starring in a Hollywood remake: 'Honey, I speed-aged the grandma!' Sixty! Surely some mistake? I'd always been precocious: a year ahead at school, first to grow boobs/French kiss/unload my virginity. Now I'd outlived my mother and qualified for a free travel pass.

It didn't help that OHW was still ten years my junior. After fifteen years, surely he should have been polite enough to catch me up. Shell-shocked on the inside, I determined not to show it on the outside. I wore a mini in pink organza with matching thigh-high stiletto boots in hot pink on *This Morning*. I was mutton dressed as lamb – the incarnation. What would you prefer – mutton dressed as mutton? Balloons and cake awaited at both radio shows. *This Morning* arranged for OHW to serenade me as a romantic birthday surprise. He was awake at that time of day – what a surprise!

That evening, I threw a picturesque dinner for my twenty closest pals on the terrace of the Bloomsbury Hotel. I shimmered in a beaded gold sheath – another triumph for the gastric bypass. My friends enfolded me in love. My girls were glowing and my sons-in-law dashing. Dan and marvellous Marc, the ravishing football translator Saskia married in 2017, were, as ever, divine. I managed to overlook the fact that OHW rolled up, utterly uninvolved, like a guest. He hadn't bought me a birthday present, not a Mars bar, not a bunch of garage-forecourt blooms. At sixty, I told myself, you're supposed to be grown-up enough not to care. Secretly, I was disappointed that amid all the congratulations flowing in from far and wide – when you're sixty you're seconds from the knacker's yard, so people make the extra effort to *shvitz* (work up a sweat) out and buy you a card – my partner managed to do precisely nothing.

A DM from Butlin's

At the beginning of 2022 I joined Instagram. I did it for work, not for fun, but took to it immediately. In just over two years I have 394,000 followers and I love every one of them dearly. Anyone on Insta can send anyone else a direct message. On a Sunday morning in November 2022 a woman who'd just seen OHW perform at Butlin's sent me a DM saying: 'So sorry about this, Vanessa. I've watched you on TV all my life, and I think if my fiancé had behaved the way yours just did, I would want to know. Your fiancé just told me he wanted to: "**** my **** and **** my ****!"'

I'll spare your blushes, Gentle Reader, and protect you from the lurid nitty-gritty. I knew she was telling the truth. The phrase she used was the exact phrase OHW used. I'd heard him say it thousands if not millions of times in the sixteen years I'd been with him. The act she described was somewhat niche and one of his favourites. This woman didn't want money or publicity. She just thought I should know. I never heard from her again.

I rang him. He screamed abuse. How could I take the word of an unknown 'clout chaser' against his? I checked. A 'clout chaser' is someone seeking fame or fortune. How could I believe a skanky whore I didn't even know? I said it was easy. She was telling the truth. He knew he'd said it. I knew he'd said it. It was his favourite phrase, in his own words, describing his favourite activity. Of course he'd said it. He said he'd go to the papers and tell them I'd ended our relationship on the say-so of a total stranger. I said it didn't matter if she was a stranger or an old friend, she was telling the truth, and if he'd asked her to do that, he must have been asking women all over the world to do the same thing, probably with considerable success. He yelled and ranted. I yelled right back. In a fortnight's time he'd be heading off to panto in Portsmouth. I said he'd better go straight there because Feltz Towers was a no-go zone.

I didn't see him, except when he came back for a few minutes to pack up essentials, for six weeks. For the first five he'd ring and howl insults at me. I hated being alone – it was every bit as lonely as I'd feared. I longed to forgive and make up, even though I knew the Butlin's lady was telling the truth. I wanted everything to be OK again. Yet OHW was so obnoxious I couldn't manage to go against my better judgement and welcome him back. Very gradually, his anger thawed. He stopped shouting and demanding the 'clout chaser's number'. (She wasn't. She didn't. I didn't have her number.)

Months earlier I'd booked to take the grandbabies to Portsmouth to see OHW as Dandini in *Cinderella*. (After years of persuasion, OHW finally took the panto plunge in 2014.) Post the Butlin's debacle I decided to cancel but didn't go through with it. Now I began to waver. Should we go? The children were looking forward to it. I had started to miss him. I knew I couldn't trust him, but I didn't have the resolve to chuck away sixteen years. I should have said in true panto spirit: 'Our relationship? It's BEHIND YOU!' Instead, I took the family to Portsmouth.

Seeing OHW was painfully awkward. It was hard to know what to say. I met the cast at the Kings Theatre. Would I, asked the theatre manager, mind appearing in the ballroom scene after the interval? The footman would announce me. I'd walk on to the stage. OHW would step forward and kiss my hand. The audience, he hoped, would applaud. I'd curtsy and exit gracefully. How could I refuse? The grandchildren, Zekey, eight, Neroli, seven, and AJ, three, were spooning in their ice cream. I said I had to slip out to make a work call. When the lights went down and the curtain rose, they were immersed in the grandeur of the palace ballroom. Imagine their surprise, when, into this sparkling dream world walked Grandma Vanessa! They'd only seen me in the ordinary world ten minutes earlier, and now I was in fairyland hobnobbing with Cinderella and Buttons. By the time I sidled back into my seat, they were flummoxed and fascinated. Had I had a good time at the ball? What was Buttons really like? How did I know the King well enough to be invited? Had I had a chance to make friends with the Fairy Godmother? I'd dismantled the wall between illusion and reality. Their love affair with the pantomime intensified. If their grandma had attended the royal ball, there was no telling what wonderments might befall us all at any moment. The Fairy Godmother might fly up the jam-and-condiments aisle at Sainsbury's.

We sat through the pantomime three times. They are bright, sensitive, musical children. By the end of the third performance, they were almost word-perfect. For weeks they lived in a pantomime world, re-enacting their favourite moments, chanting catchphrases, singing the songs. Gradually my anger abated. Like the children, I was transported by make-believe. I allowed myself to be soothed and reassured. I didn't want to be alone. I feared the cruel rupture of a break-up. I wanted everything to be all right.

I am so sorry

I regret taking my grandchildren to that pantomime with every fibre of my being. It was a terrible mistake and I take full responsibility for making it. Instead of calling time on the relationship, I allowed them to see OHW in an enchanted sphere, further cementing their love for him. I am so sorry.

I flew to beloved East Cork with my family for Christmas. OHW was still in panto. He phoned daily. 'Only fourteen nights till we are back together. Miss you. Love you.' I hoped he meant it. Christmas in Ballycotton is heavenly. Every day brings all weathers. We bundle up against the rain with sunglasses perched on our anorak hoods. We have learned to expect a rainbow followed within the hour by burnished sunshine, torrential rain, hail, sun, wind, gloom, fog. We dig to Australia on the golden, often empty, beaches. We feast at the array of cafés and restaurants in what is accurately called the 'larder of Ireland'. My son-in-law Dan joins the local all-year-round Atlantic swim and fishes for mackerel in the harbour. We have a wonderful time. Santa made a crowd-pleasing appearance on Christmas Day.

71

A tip-off to my daughter

SASKIA LOOKED TROUBLED. SHE WAS breast-feeding new baby Cecily. 'Darling, are you all right?' Someone who knew someone who knew someone had tipped her off that OHW was involved with another woman. The person knew OHW and referred to my having 'thrown him out' before panto. I hadn't told anyone. OHW must have told this informant.

Saskia, Allegra and their husbands were in the loop. I phoned OHW. 'Saskia has been told you are having an affair. It's clear the person knows you and your movements.'

'Just some nutter,' he said calmly.

'But they seem to know you and what's gone on between us.'

'I drink, sweets. I drink heavily. When I'm drunk, I get loose-lipped. It could be anyone. Forget it. Have a lovely holiday. Only nine days till we're back together.'

I did my best to forget it. It seemed feasible OHW could have blurted out information to some passing stranger in a pub. Move on, Vanessa, don't dwell on it.

OHW was back at home. I was back at work. He seemed fractious, short-fused. After the camaraderie of panto, he was out of work and out of sorts. On 12 January I told him I was off to Hamleys, the toy shop in Regent Street, to buy a present for my grandson Zekey's ninth birthday. OHW said he'd come along and buy a gift for Zekey too. He stopped at Molton Brown and bought himself an expensive deodorant. Mooching about the shop, I overheard him talking to the assistant about having something engraved. I moved away sharpish, assuming he'd bought me some perfume to make up for the lack of a sixtieth birthday or Christmas present and was arranging to have a message engraved on it.

At Hamleys he chose an expensive robot for Zekey. As I queued up to pay, he disappeared. As usual, I didn't know what to do. If I made a fuss and said, 'You told me you wanted to buy a present. That means paying for it,' he'd give me hell, loudly in public. I could afford to pay for both presents. Doing so would avoid a collision. I paid up.

He stopped at Molton Brown and came out with a package. I waited for him to give it to me. We walked up Regent Street in silence. I couldn't stop myself asking: 'What did you just pick up?'

'An aftershave.'

'Did you get it engraved?'

'Yeah.'

'Who for?'

'Myself.'

'WHAT?'

'Yeah. My name. What the fuck is wrong with that?'

'You bought aftershave, and had it engraved with your own name? Don't you think that's unusual? I thought it might be a present for me.'

'Why don't you just fuck off?' he was screaming. Passersby were staring. Any minute any of them might whip out their phone and send the footage to *MailOnline*. 'You can't just let me live. You've always got something to say. Shut your mouth. Shut the fuck up.' He was bellowing, snarling. I was trying to soothe him, wondering why I'd let him back into my home.

OHW said he was going to a family meeting on Sunday. On Saturday I asked: 'What time are you meeting your family tomorrow?' I had no agenda. Whatever his answer, I was fine with it. I just wondered if I'd have time to pop over and see the girls, or if I should arrange lunch with a friend.

A surge of livid lava cascaded out of him. 'YOU WANT ME TO TELL YOU WHAT TIME I'M SEEING MY FUCKING FAMILY? How dare you? I don't have to be strapped to you twenty-four hours a fucking day. You do you. Do you fucking understand me? You won't see me, yeah. Fuck off. What time am I seeing my family? Fuck yourself. If you see me, yeah, don't talk to me, yeah. How fucking dare you?' On and on went the yelling. I knew exactly what I'd said and how I'd said it. I couldn't work out what was wrong. I retreated to my bedroom and locked the door. He was on the landing, still yelling.

Hours later, I had to come out. I was taking the family to the Theatre Cafe Diner for Zekey's birthday. I was looking forward to seeing my grandchildren's expressions when the waiters and waitresses sprang on to the tables and belted out songs from the musicals. OHW hailed a taxi. He said: 'Why are you frowning up your face?'

The answer was: 'Because I've been locked in my bedroom for four hours because you yelled at me for saying something perfectly normal and I'm upset.' I couldn't say that. He'd go berserk. I couldn't ignore the question, he'd go berserk. I couldn't say: 'Frowning, darling? You must be mistaken. I am smiling beatifically.' He'd go berserk. I squirmed and tried to make conversation with the driver.

When we arrived, the children were already bopping away at the table. OHW got stuck into the Malbec. Earlier in the week, as a concession, he said I could join him at a birthday celebration somewhere in South London later. Burgers were demolished. We were convinced our waiter's 'Let It Go' was head and shoulders above the others. Dessert appeared.

OHW growled: 'The taxi's here.' I was startled. We hadn't finished. I was still eating. I hadn't paid the bill. More importantly, I hadn't had a chance to kiss my grandchildren goodbye. 'Please, hold on a second. Let me just sort things out.' I stood up, trying to put my coat on, gesture to the waiter, kiss the baby, say goodbye. I felt pressured to clear out fast for the waiting taxi and flustered attempting to do too many things at once.

Saskia called me over. 'Mum, I'm so sorry. I've been given more information. OHW has definitely seen this woman since Christmas. Allegra and I think you need to know.'

OHW denied it. He muttered: 'Just some nutter. Never heard of them. Ignore it.' He turned to walk out of the restaurant.

Saskia called him back. 'You do know this person,' she said, almost whispering so the children wouldn't overhear. 'You know exactly who this is.'

Allegra spoke quietly but firmly. 'You are a liar.'

In the taxi, my daughters' gentle voices replayed again and again in my brain. *You do know who it is. You are a liar. You do know who it is. You are a liar.* My daughters had finally spoken out. The informant had involved Saskia. If it had been me, I might have tried to cover up the incident as I had in 2017. This time my family were implicated, and I didn't have the option of pretending it hadn't happened.

I asked him directly: 'Who is it? Who have you been seeing? You do know. You know exactly who this is. Please tell me. What is happening?'

He was speaking but the sounds coming out of his mouth were from no language I have ever heard. He was spluttering, stuttering, expostulating. I didn't understand because he wasn't making sense.

He shouted a woman's name. 'Cathy!' I'd heard all about Cathy, his first real girlfriend thirty-two years earlier.

'Cathy? What do you mean? You haven't seen her for thirty years.' A jumble of thoughts clanged and clashed in my head. Cathy? Had he been seeing her all along, having an affair throughout our sixteen years together? OHW and Cathy must still be together. Our relationship had been built on a pack of lies.

He was still speaking in tongues. Nonsensical syllables poured from his mouth. I couldn't stand another second. I couldn't bear it. I opened the door of the moving taxi and jumped. My coat was torn. Blood was seeping from a cut on my arm. It was dark and cold. I was in a back-street somewhere in South London. I assumed OHW would stop the cab and follow to see if I'd survived the leap. He didn't. I HAVE NOT SEEN HIM SINCE.

Shaking, dabbing at the blood, shivering, wondering how the hell to get home, I looked at my phone. OHW was sending message after message. They were not meant for me. He thought he was messaging Cathy. When he realized I could see them, he deleted the messages.

He went to the party and posted pictures on Instagram of himself drinking champagne and posing. When I eventually reached home, I crawled around outside my back door scrabbling in the mud to find OHW's door key. He couldn't trust himself to look after a bunch of keys when drunk – he'd lost dozens – so he'd taken to burying them. I found it, went inside, locked up and sent him a message. 'I have your key. You will not be able to get in. I have gone to a hotel. There is no point coming back. I won't be there. Do not come back.'

I wondered if there might be any clues to what was going on. I went to look in his dressing room. 'What do you think you'll find?' I asked myself. 'A monogrammed handkerchief? What do you think this is, you idiot – *Othello*?' In the first drawer I opened

I found a receipt, dated just before Christmas, from Pandora the jewellery company. The purchases: a letter L, alphabet charm, a January birthstone, small O pendant and cable chain necklace. Total price £109. In the morning, I sent a picture of the receipt to OHW. He texted back 'Secret Santa.'

By text much later the next day he admitted the woman wasn't Cathy. That had been a lie. She was someone he'd been sexting/befriending/counselling/any other verb you like for a year. He was scared she'd tell me. She was naked. There were pictures.

My family and I packed twenty-five suitcases, bags and bin bags with OHW's possessions. We counted a hundred pairs of trousers. Someone in a van picked them up on Friday, 20 January 2023. I was at work when his worldly goods were collected.

72

Pastures new

A BOUQUET OF DOOM ARRIVED from Radio 2. It was a motheaten affair. Think seventies Interflora on a bad day. Can a bunch of flowers be insulting? Without a doubt. Controller Helen Thomas was viciously culling. The days when adored presenters David Jacobs and Desmond Carrington died in the saddle and Jimmy Young presided over the lunchtime slot until they binned him at the age of eighty-one were over. These were Sir Jimmy's last words on Radio 2: 'Just so we're all singing from the same hymn book, it was not my idea to go. I didn't want to leave you at all, and I know from your messages you didn't want me to go either.'

Radio 2 was trying to recruit a younger audience, seemingly without caring about the devoted listeners who'd been loyal to the station for decades. Flinging beloved presenters on the scrapheap, despite their proven popularity, is sticking two fingers up to the audience. 'You love Presenter A? You listen to him every day and set your clock around his show. We couldn't care less. We don't want you Old Fogeys listening. Yes, even though we are the BBC and don't need to sell advertising. You'll never hear your favourite presenter here again. Sod off.'

Simon Mayo, squeezed out of his own show after being forced

to 'share' with Jo Whiley the programme he'd hosted solo for eight years, packed his bags. Graham Norton headed for the hills. Craig Charles was given the heave-ho.

Unconscionably and incomprehensibly my dear friend, radio titan Steve Wright, was removed from his afternoon show, despite stellar listening figures. Listeners were outraged. They'd grown up with Steve. We all had. He'd been our friend on the radio for forty entertaining, unpredictable, enjoyable years. Steve, devastated and trying hard to keep a lid on his heartbreak, told me personally he was turfed out of Radio 2 with no show and no job and had to ask for a meeting with Director General Tim Davie to request (he didn't use the word beg) to keep on presenting *Sunday Love Songs*. Tim Davie agreed Steve, who'd been at the Beeb forty-two years, had been treated shamefully and overruled the decision. Steve retained *Love Songs* but lost his raison d'être. Stripped of his daily relationship with his audience, his gallery of guests, beloved posse and the outlet for his cauldron of creativity, he struggled to be positive. Along with his fans and friends all over the world, I was sad beyond words when he died on 12 February 2024. His last text to me said: 'Hi hi V. You are a gr8 broadcaster, a beautiful person. And I am a fan. Wishing u only the best xx steve.'

The cull continued. To my horror Paul O'Grady, indisputably a radio genius, with an instantly recognizable voice, original turn of phrase and caustic line in sardonic humour no one in the world will ever be able to emulate, was told there was no longer a place for him at Radio 2. The decision was ludicrous. Paul had never been more popular. He was beloved and irreplaceable. He told me he was privately bereft. I'd loved Paul ever since I inherited his cast-off frocks when I succeeded him on the *Big Breakfast* bed. To my blushing delight, he called me the 'Delphic Sibyl'. I combed my vocabulary for effulgent epithets to bestow on him. I loved and admired him boundlessly. His premature death at just sixty-seven was a body blow to millions. He was worshipped the length and breadth of Britain. His kindliness illuminated all he encompassed. His inspired producer and kindred spirit Malcolm Prince, featured frequently in mock-exasperation on Paul's programme, remains my friend and mentor. If Radio 2 no longer considered Paul O'Grady an integral star in their constellation, I was sure they'd be marking my card and I'd be next to walk the plank.

Jump? Or wait to be pushed? I know which I prefer. I'd been circled by poachers from rival radio stations for years. When Rupert Murdoch bowled up – OK not actual Rupe, his representatives on earth – waving a bulging cheque book, for the first time in twenty years I sat up and took notice. Radio 2 was signing up presenters younger than my best brassiere. Local radio teetered on the brink of butchery. How long would it be till Auntie delivered on the looming threat to pull funding and amalgamate local stations? The BBC would knowingly be delivering the fatal blow to local broadcasting. Local radio must be local. The minute you blend local stations to save cash you haemorrhage USP and kill the concept. If you live in London, no matter how diverting BBC Radio Kent might be, you have no interest in tuning in. BBC management knows perfectly well that combining local radio stations sounds their death knell. They pretend they don't realize, while behind the scenes they prepare to throw up their hands in mock-horror when audiences melt away. As they switch off the lights at local radio stations, they'll shake their heads sadly saying: 'Such a shame. No one listened anyway.'

I didn't fancy (a) getting the Big E from Radio 2, or (b) clinging to the sinking raft of BBC Radio London as funding ebbed away. The clincher was mistakenly booking holiday on the very day of the Platinum Jubilee celebrations. My daughters noticed. 'You'd better cancel that trip, Mum. You know they'll never let you be absent on such a huge occasion.' Historically, they were right. Bosses always made sure I was around for the big ones. I'd be instructed months in advance not to book leave on the day of mayoral or general elections, royal weddings, the London Olympics, Notting Hill Carnival ... Glimpsing the Boss, I said: 'I've booked to be away for the Platty Joobs. I suppose I'd better unbook.'

'Oh, no,' came the reply. 'Have a great time. Enjoy!' That would have been very civilized if it hadn't signalled something significantly awry: 'We are the BBC's designated *Breakfast Show* for London and we have nothing whatsoever organized to mark the Queen's jubilee.' I'd had a reserved seat at the Palace press cavalcade to report live from the Golden Jubilee, a double-decker bus in the middle of Hyde Park with a ladder to climb on to the roof for Kate and William's wedding, and a berth on the royal shallop *Jubilant* for Boris's Thames Pageant for the Queen's Diamond Jubilee. In 2022 there was nothing. I decided to accept Mr Murdoch's generous offer.

I leave News UK on Friday at six p.m. And sign to Global on Tuesday at six p.m.

Did you watch *Succession*? Fact: I was at News UK for twenty months. You fill in the rest. My last Talk TV show finished at six p.m. on Friday, 26 April 2024. On Monday, 29 April I was secreted in an SUV beneath Global's headquarters in Leicester Square, then smuggled upstairs to a room with brown paper taped over the windows, where Hair, Make-up and Global's press and publicity team awaited. Photos were taken, then energetically retouched. On Tuesday, 30 April I signed the contract. On Wednesday the thirty-first, in a shocking-pink Michael Kors suit trimmed with marabou feathers, I sailed in to my chum broadcasting legend

Nick Ferrari's studio to announce my Saturday LBC three to six p.m. show. Global moves fast.

I was in full flow, headphones on, shoes off, just in time for breaking news of Sadiq Khan's historic election to a third term as Mayor of London. LBC's courting dance might have taken the best part of twenty years, but I am jubilant to have been wooed and won. Is it absurd to hope, at sixty-two, that my best broadcasting years are yet to come? OK, it *is* absurd. I hope so nevertheless.

73

I've been out every night for 530 nights

AT THE TIME OF WRITING this, I have been out every night – except one – for 530 nights. You think I must be exaggerating. I'm not. Every single evening since my relationship imploded on 14 January 2023, I've trekked off somewhere, anywhere, so I wouldn't have to be stuck inside all alone staring at the wallpaper. There's nothing wrong with the wallpaper, by the way. It isn't Lulu Lytle, but it's a pedigree Sanderson floral. There's nothing wrong with the house either. It doesn't creep me out. I don't rattle around wretchedly in my empty nest. It can even veer on cosy when the fire is lit. It's not the house's fault I haven't wanted to sit in it.

I sallied forth because to subside on to the sofa you need to be tranquil. Watching telly requires no effort unless being solo makes your heart thump and your mind race. Ditto relaxing in the tub. Serenity breeds contentment. Pre-heartbreak, schlepping off the premises just for the sake of it was my idea of earthly torment. Why don crampons and sign up sherpas for an onslaught on the West End when you can munch Welsh rarebit comfortably at home? I'm as partial to a foray on a red carpet as the next TV personality, but I picked and chose. Occasionally, envelopes were opened in my absence.

Post break-up, though, everything changed. I was sad to be

partnerless. Ever since my hideously well-documented 2000 divorce, I've been battling the psychological scars of unwanted singledom. As you know, signed, sealed and delivered to the doctor Grandma Sybil chose for me in 1985, I naively believed I'd been released from the stigma of singledom for ever. I certainly wasn't prepared for another break-up. Weeks before my sixty-first birthday I was suddenly solo yet again and I couldn't hack it. Sit tight, face your fears, Netflix binge, take up weaving, cook yourself a casserole. Fat chance. The idea of hanging about all by myself till I left for work the next day – twenty hours later – was too echoingly empty to contemplate. So, I didn't do it. I went out instead. Out-out is sociable, stimulating and bristling with promise. Mr Right won't accost you within your own four walls unless he's bringing the Deliveroo. Outside them, there's a slim chance of an encounter, and failing that, you've had a laugh, eaten dinner you haven't cooked and passed the time until you've reached a civilized hour to toss and turn sleeplessly till morning.

Where have I been for 530 nights, for Heaven's sake? I've whirled about with school friends, work friends, old friends, friends I wasn't sure were friends, colleagues who'd wanted to be friends but never dared ask and people who'd floated about the periphery of my life who, it turned out, were prepared to stand up and be counted. My best heterosexual platonic friend Marcel Knobil – we haven't, he doesn't want to, I don't want to, we love each other, eternally adorable chum Tina Cole (and her delectable husband Mike), confidante Fiona Phillips (and her charmingly challenging husband Gary), beloved school friend Alex Katz (and her erudite husband Andy), infinitely hospitable Laura and Ian Langdon, former boss and pal Gloria Abramoff, holiday companion Niki de Metz, close cousin Dalya Shear, wise, warm, inspirational Linda and Mark Reuben, old flame Danny Lachter, my staunchly loyal and inspirational senior producer and pal

Mark Machado and a smorgasbord of compadres have buoyed me up and soothed me down. Cheerleaders Beverley and Howard Calvert, cousins-on-both-sides Beverley and Nicholas Crowne and their wondrous son Benjamin. Showbiz pals have been tremendous too. Myleene Klass, Claire Sweeney, Linda Robson, Holly Willoughby, Rylan, Denise van Outen, Alison Hammond, Sara Cox, Elizabeth Day, Nicky Chapman, Liz Earle, Sharon Marshall and a cocoon of others took me to screenings and launches and made me Marmite sandwiches in their kitchens. Throughout his considerable trials and tribulations, Phillip Schofield has been immensely supportive. I took refuge in beloved East Cork with family, swept my exhausted grandchildren to pantos, Sunday-morning kids' movie premieres and out to dinner in Hendon whether they wanted to go or not.

Celebs Go Dating

Yes, I've dated, on telly, somewhat impatiently in *Celebs Go Dating*. I adore the show, but it's tailored for the young and buff. If you're a *CGD* oldie the odds are stacked against you. Think about it. Nubile twenty-somethings sign up because they want a springboard to *Love Island*. No sixty-plus bloke worth his salt wants to risk public rejection by a celebrity on a reality dating show. If you're still a wannabe at sixty-odd there must be a whiff of iffiness about you. Producers and researchers admitted the pickings in the senior age bracket are slim.

If you watched the show, I want you to know I didn't really stomp out while that monumental mitherer from Manchester was mid-munch. I was unexpectedly called in to report on a breaking news story at work. The show's producer interrupted the date to tell me I had to dash off. I politely explained why I had to make a sudden departure. My courteous explanation ended up on the cutting-room floor. I agree, my behaviour looked terrible. When I watched the episode, I was horrified. Believe me, I'd never be that rude in real life and I sure as heck wouldn't be so blatantly discourteous with six TV cameras whirring. You might not believe me, of course. You know what you saw with your own two eyes, and you saw me up and off without a backward glance. It's pathetic to sign up to reality TV and blame the edit. It's been done so often viewers never buy it. All I can do is swear on a stack of Bibles I didn't stalk out leaving the fellow high and dry and reassure you if I had dinner with you, you'd be able to confirm I'm a charmingly mannered pussycat.

Real-life dating

Off the telly I've dipped a pedicured toe into the dating scene but only through personal introductions from friends and friends of friends. I can't risk using an app. Whatever happened would be on page five of the *Sun* the following morning and I couldn't face it. I haven't had much luck so far. First was the extremely Orthodox – synagogue seven mornings a week, kosher to the max – chap with an unexpected penchant for nightclubs who thought he was a dead ringer for Al Pacino. He certainly used the same ebony hair dye as Scarface. Four months of wining, dining and writhing about in the boudoir led me to the are-we-an-item? conversation.

'Of course!' he answered. 'We're going out four times next week and to the South of France in the summer. Of course, we're an item.'

There were drawbacks. His favourite question was 'Am I the handsomest man in the room?'

I'd say, 'Definitely!' and wait for him to reciprocate with something complimentary about me. I'd wait. And wait. He never bothered.

He went to a wedding in the USA, rang me every day, phoned from the carousel when he landed at Heathrow and took me out that night. Two days later, on a sun-drenched Sunday, he came with me to a dear friend's party. He sauntered off in search of a cocktail and I greeted my hostess's brother who'd been at the same US nuptials. 'Hi, sweetheart. Did you have fun at the wedding?'

The brother turned aubergine-purple and started spluttering. 'I can't talk about it. I can't talk about it. You'll have to ask my wife.'

I trotted over to the wife. 'Did something happen at the wedding?' Gentle Reader, I'll spare you the details. Turned out Mr Religious boasted to my chum's bro he'd spent a 'wild night' with one of the female guests. The brother and his wife said they couldn't believe Mr Religious 'had the audacity' to come to the party with me. He'd told everyone at the wedding he was single. I was banjaxed. I'd fallen for the observant God-fearing butter-wouldn't-melt routine.

I went over to him and said quietly: 'Did you have sex with a woman in America?'

He replied: 'I admit it.'

I waited for an excuse or explanation: *I was drunk. I was on ketamine. I'm a sex addict. I didn't enjoy it. I was thinking of you all the time.* He didn't say a word. 'Oh, my God!' I said. 'I'm so hurt. I'm so shocked. Oh, no. I can't believe this.'

He managed one grudging 'Sorry', followed by 'Now put some make-up on, go in and enjoy the party.'

Since then, I have (mostly) avoided being up close and personal with (1) the well-known sports agent who 'took me up the Shard' and whispered seductively, 'I'm going to kiss you all the way down in the lift.' Jeezus! Thirty interminable floors of drool, saliva and fat slobbery tongue rotating like a second-hand Zanussi. Never have I been so relieved to hit dry land; (2) the entrepreneur who tipped off the paparazzi we were having dinner and looked livid when I spotted them, ducked into the Ladies and refused to come out till he'd vacated the premises; (3) the property developer who – I kid you not – insisted on showing me a photograph of his dead mother, surely an appalling infringement of first-date etiquette; (4) the fellow – someone's cousin's neighbour, no one admitted to knowing him personally – so ancient he looked as if he'd been exhumed that morning – I kept thinking clods of fresh earth were going to fall into the soup; (5) the insufferably pompous legal beagle with a Roller and handsome chauffeur. Did I manage, through lingering looks in the rear-view mirror to arrange a post-date assignation with the handsome chauffeur? Affirmative; (6) the chap I'd chucked in 1974. Forty-nine years later he was still a pill. An ex, even if you were twelve when you gave him the heave-ho, is always an ex for a reason. You get the picture.

Cupid may be playing hard to get, but the quest for romance hasn't been unremitting gloom. There've been more than a few close encounters of the non-nauseating kind and a handful of suitors I'd happily take home to Mother if my mother hadn't been dead since 1995. The 'slap and tickle' issue is hilarious. Obviously, I'm not a virgin. I'm a grandma of four. Yet I'm not comfortable leaping aboard a stranger's todger on a first date. I like him to compose at least one decent couplet to my eyebrows first.

Sunday, 28 January 2024 has been my sole night in so far. Frankly, I was too knackered to budge. I cooked a salmon (shame it wasn't sole) fillet in foil in the oven. I watched something so forgettable on TV I can't remember a frame of it. I wallowed in scented bubbles. I didn't discombobulate. It was bearable. I survived twenty hours of my own company intact. I wonder if I might even try it again. I'm writing this on 25 June and I haven't so far, but you never know.

Epilogue

THIS BOOK IS OVER 90,000 words long and it is time I flitted through the loves of my life. Don't worry, I'm not going to drone on about symphonies, sonnets and soufflés or shove a raft of literary recommendations down your throat. The list is short but sweet: my girls, my grandbabies, my chums, roses, peonies, hellebores, snowdrops – did you know someone who loves snowdrops is called a galanthophile? – polysyllabic words, especially 'galanthophile', rooms with verdant views curtained in rose-strewn glazed chintz with matching rose-strewn wallpaper and petit-point cushions, films in which beautifully dressed people converse intensely to an occasional jazz accompaniment while nothing happens: Woody Allen's *Everyone Says I Love You* (1996), Bob Fosse's *All That Jazz* (1979), George Cukor's *The Philadelphia Story* (1940).

I abhor being asked what my favourite anything is. Become a celeb and they ask you that question dozens of times a day. Five favourite things to do on a Sunday morning. Five fridge leftovers to recycle into dinner. Five supersonic sex positions. My brain shuts down at the sheer violence of the edit. When pressed, I can name my favourite film: Alan Parker's *The Commitments* (1991). Dublin, Roddy Doyle's humour, sweet soul music and dreamer Jimmy Rabbitte's one-sided 'dialogue' with my beloved mentor and avuncular friend, the late Sir Terry Wogan, collide in two hours of unalloyed bliss.

I love a lyric. I don't mind if it's Dorothy Parker's 'One Perfect Rose' in which she scoffs at flowers and laments the fact no man has ever bought her a 'perfect limousine' or Dylan Thomas's 'Under Milk Wood' with its 'fishingboat-bobbing' sea, Cole Porter, Tim Rice, my late great friend Benjamin Zephaniah – I'll never forget his dreadlocks sweeping across my living room floor as he followed Mr Kabs from Flawless's choreography at my 58th birthday party or T. S. Eliot. My fondness for a felicitous phrase means I have frequently been felled by a fellow's fluency when I should have seen through the alliteration to the vacuum within.

I love my daughters. Allegra and Saskia are magnificent women, courageous and public-spirited, fun and kind. My darling Myleene Klass calls them queens. If we weren't related, I'd be honoured to be their friends. I love the way they love and support one another. It is balm to a mother's soul to see her adult children holding hands. They are gentle, patient, imaginative mothers. Watching the compassionate way they nurture their children – reading them every unutterably tedious word of *Swallows and Amazons* and its ever-lasting prequels and sequels, allowing them to spatter the kitchen with cake mixture, reciting poems I recited to them when they were little till they fall asleep and singing *Joseph and the Amazing Technicolor Dreamcoat* backwards, forwards and sideways, while wiping up fingerpaints, assisting with a project on 'wind turbines', teaching them to thin sweet peas and enforcing zero 'screentime', I dare to hope they might have learned some snippets of maternal magic at my knee. We are lucky. We glory in one another's company. We rarely have a cross word because they don't make me cross. I cherish and relish both. Ideally, I'd copy the Ewings of *Dallas* and buy a ranch like Southfork where we could all live together.

I've done the next best thing. In 2017 I blended my best people with my best place – Ballycotton in God's own East Cork. I had ideas above my station and bought a house on the Wild Atlantic Way. Get me! Vanessa Two Houses. I don't come from a two-home kind of family. After an adrenaline surge when I signed on the dotted line, my spirits plummeted. What if I couldn't afford it? What if we never used it? What was the place like in the dead of winter, for Heaven's sake? I had no way of knowing we would all fall head over heels in love and spend every possible moment we can snatch away from work in anoraks trying to stop the babies being blown away on Ballynamona beach. I called the house 'Ballykneidel'. 'Bally' comes from the Gaelic *baile na* meaning 'place of'. *Kneidel* is the Yiddish word for a matza ball. Sometimes we receive post addressed to Mr Barry Kneidel.

How does a Jewish girl from North-west London come to leave her heart on the island of Ireland? Blame a gynaecologist's waiting room in 1994 where I read about the world's most wonderful country-house hotel, Ballymaloe, and somehow persuaded my then husband we absolutely must visit. Chatelaine, the late Myrtle Allen, mother of modern Irish cuisine, was in her prime, feeding guests home-grown ambrosia. A bite of Myrtle's soda bread was celestial. Spread the slice with freshly churned butter and you were in sensual nirvana. Add a dollop of homemade gooseberry jam and you vowed never to touch mere Hartley's again.

Is it acceptable to say one of the most passionate love affairs of your life began with Myrtle's apple muesli? Ballymaloe became the place the girls and I dreamed of when we were sick at heart. Wandering Jews don't have Daphne du Maurier's Manderley. Ballymaloe became our symbolic sanctuary. Owning Ballykneidel meant exploring, and instantly East Cork transmogrified into 'beloved East Cork'. When we fly into Cork airport, we can never believe the posters saying: 'Fly Direct from Cork to Valencia/Amsterdam/Venice.' We don't want to. We have flown to Cork. That is our number-one destination. Why would we want to be anywhere else? We drive out of the airport and turn left. Not right. We are firmly East not West. Occasionally we make sorties to Kinsale and beyond, but only under cover of darkness.

I love motorbikes. OK, I can't drive one, but I can be whisked around the world on the back of the supersonic London Limobikes currently breaking the sound barrier on a street near you. Think about it. Stew in a car or cab stuck in stagnant traffic inhaling the driver's armpits, or vault aboard a lean, mean, purring machine and indulge in light biker-chick badinage via a clever in-built helmet contraption with your ex-motorbike racing champion driver? If I can I rev up everywhere on a Yamaha FJR 1300 I do. Full-length evening gown? Bundle it up and leap. Stilettos? Ditto. I roared up to Rupert Murdoch's summer party and dismounted – in leathers – much to the consternation of Rishi Sunak.

Did I mention I love dresses? The minute I exploded on Instagram I noticed nine out of ten messages – rather impressive dick pics apart – were 'Vanessa, your dress! I love it. Where can I find it?' There was clearly a gap in the market for elegant, easy-to-wear, feminine, life-affirming, colourful dresses for women who might have reached a certain age and begun to feel just a soupçon invisible. The *schmatta* business (rag trade) is in my DNA. I knew if I could manufacture a capsule range of no-zip, no-fasteners, machine-friendly, crease-free, exquisitely cut, flattering, cleverly designed dream dresses they would sell. Luckily my friend Katrina Shalit made sure my stars collided with Linda – imposingly incisive judge on Lord Sugar's *The Apprentice* – Plant's. 4love.uk was born on 4 December 2023 and the dresses – plus separates, evening gowns and, *noch* (would you believe), vegan-fur gilets – have graced the covers of innumerable magazines and made thousands of women feel fabulous. Extra delight cascades when they send me stunning photos looking sumptuous in my dresses and having a ball. Am I elated to have been part of the launch of a thriving enterprise in my seventh decade? I'm tickled to death and both my grandpas would be proud.

I've left the loveliest till last. Ladies and gentlemen, may I present my grandchildren Zekey (Ezekiel Jack), ten; Neroli Joy, eight; AJ (Amiel Jeffrey), five; and Cecily Violet, twenty months? I say 'may I' but you'd have a hard time stopping me. Obviously, they are all remarkably good-looking geniuses. That goes without saying. Even better, they are the Blackpool Illuminations of my life.

When Zekey was born, I told an interviewer I was overflowing with so many hormones I was sure, if I really tried, I could breastfeed him. He is musical, scientific, well-read, laughed so ebulliently at the opening night of *Fawlty Towers* at the Apollo that Michael Palin said the sound of his giggles was a tonic, and he is a fourth-generation Haberdasher. In June 2024 Zekey wrote an essay about me. Allegra forwarded me the final sentence: 'So that is the life of Vanessa Feltz so far. Her heart is full of generous things to do and kind things to say.'

Neroli Joy is nominative determinism personified. She is as fragrant as the seventeenth-century Princess of Nerola who gave her name to the essential oil of orange blossom she discovered, and is a smiling, giggling harbinger of happiness. Neroli devours several books simultaneously, designs and makes her own clothes on a sewing machine – my great-grandpa Feltz the tailor would approve. She adorns North London Collegiate School where she won the poetry prize and sent me this letter in the post: 'Nobody has someone like you. I love you. If I lose you, I will cry for the rest of my life. THE END P.S. Love you Grandma.'

AJ oozes charisma. He is the Maurice Chevalier of Mill Hill. At five years old AJ is an infant Stephen Fry, comical polymath, bon viveur, master baker, climber of sheer walls and performer of a vast repertoire of songs from all eras. He teases me mercilessly, insisting I am a baby because I like fruit puree. He is an incandescent star on my Instagram.

Which brings me to Cecily Violet, the baby who refuses to be a baby. She is tiny, about the size of a Friday-night chicken but for Ceci twenty months is the new twenty-five. She is so adamant she's a grown-up lady she manages to convince the world. She doesn't just walk, she strolls. She doesn't merely talk, she converses. She has views, opinions and a take on current affairs. 'Saskia, have you remembered the D-U-M?' I asked.

'Daddy's pocket!' piped up Cecily. Her sentences sizzle. Her vocabulary is so abundant it's almost alarming.

Carrying this tiny talker down the Edgware Road, I heard a passerby ask in wonderment, 'What did that baby say?'

History repeating itself. Nature and nurture. Isn't that where this book began, sixty-two years ago with a stranger asking my mother, 'What did that baby say?'

Acknowledgements

Many thanks to my daughters Allegra and Saskia, sons-in-law Dan and Marc for the technical back-up, Cosmo Landesman for his insightful support, Eugenie Furniss for getting the thing off the ground and all the friends who smiled wanly when I ambushed them with paragraphs. Susanna Wadeson and Stephanie Duncan at Penguin Random House have done a bang-up job.

Index

Abramoff, Gloria 76, 323
Adams, Kaye 106
Agassi, Andre 81
Airport (BBC TV) 179
Aladdin 244–5
Alagiah, George 153
Alexiou, Douglas 216
Alkabetz, Shlomo 219
All That Jazz (film) 328
Allen, Myrtle 330
Allen, Woody 328
Allsop, Malcolm 106, 109, 110, 111, 112, 115, 118, 173, 197
Alves, Dulce 283
Andre, Peter 178, 179
Andrew, Prince 77
Andrews, Anthony 39
'Angel of the North' 207, 228, 229–31, 239, 265, 269
Anglia Television (now ITV Anglia) 106–8, 112–16, 118, 119, 131, 183, 194, 197
Animal Hospital (BBC TV) 154
Anisfeld, Irene 26
Anisfeld, Marcel 25, 26
Anisfeld, Sharon 25, 128

Anka, Paul 167
Anne, Princess 89, 92
Ant(hony McPartlin) and Dec(lan Donnelly) 274
Apprentice, The (BBC TV) 331
aufruf 89–90
Aznavour, Charles 143

Baker, Danny 251, 252
Balfour-Lynn, Dr Lionel 11–12
Ball, Zoe 136
Ballycotton, East Cork, Ireland 310, 324, 329–31
Ballykneidel 330
Ballymaloe House Hotel, East Cork, Ireland 330
Bannister, Matthew 202
bariatric surgery 277–80, 290–92
barmitzvahs 21–2, 189, 270
Barrymore, Michael 192
Basil Brush 83
Bateman, Nick 232
batmitzvahs 189–90, 227, 246, 247, 285
Bay City Rollers 136
Bazalgette, Peter 254

BBC Radio London 250, 251–62, 282, 319
   The Breakfast Show 251, 252, 261, 287, 295, 300, 320
   Jewish London 74, 75, 127, 250
BBC Radio 2 163, 254, 317, 318, 319
   The Early Breakfast Show 281, 282, 287, 289, 296
   Sunday Love Songs 318
BBC Television 2, 3, 76, 80, 146, 153, 170, 178, 192, 297, 300, 319
   see also specific shows
Beatrice, Princess 77
Beckham, David 257
Beckman, Sam 53
Beckwith, Tamara 242
Bee Gees, the 281
Benitah, Allegra 329
  batmitzvah 189–91, 246
  birth 63
  childhood 67, 102, 108, 118, 124–5, 131, 141, 147, 161, 167, 175, 179, 189, 226–7, 245
  children 152–3, 329, 332
  and 'One Hit Wonder' 311, 314
  and parental divorce 205–6, 208–9, 211, 212, 214, 215, 219
  and Saskia 64–5
  school 126, 129, 141
  and Vanessa and Platinum Jubilee celebrations 319–20
  and Vanessa's appearance on *Celebrity Big Brother* 238, 239
  at Vanessa's 60th birthday celebrations 307
  wedding 284–5
Benitah, Dan 284, 307, 310, 311
Benitah, Ezekiel Jack (Zekey) 308–10, 312, 313, 324, 332
Benitah, Neroli Joy 308–10, 313, 324, 332
Benson, George 143
Bermant, Chaim 70, 87
Beverley Sisters 69
Big Breakfast (Channel 4) 4, 134–5, 136–8, 139–44, 145–9, 150–54, 155, 163, 173, 174, 179–80, 183, 189, 192, 318
Big Brother's Big Mouth (Channel 4) 253
Birds Eye advertisement 155–7
Black, Cilla 248
Blackwell, Fred 69
Blethyn, Brenda 163
Bloom, Helen 8, 130
Blue (band) 270
Blumenow, Stephen 192
Botton, Alain de 276
Bowers, Dane 270
Bowie, David 281
Boy George 155, 160
Boyzone 234, 237
Brady, Baroness Karren 155
Branagh, Kenneth 40
Brand, Russell 253–4
Brecker, Richard 74
Bremner, Rory 288

Britton, Fern 277–8
Bruce, Fiona 153
Bruce, Ken 281
Bryan, John 77
Burke, Kathy 180
Bussell, Darcey 287
Bygraves, Max 180

Calvert, Beverley 324
Calvert, Howard 324
Campbell, Glen 179–80
Carlisle, Belinda 100
Carlyle, Robert 180
Carnie, Lewis 281
Carrey, Jim 136
Carrington, Desmond 317
Carroll, Sue 214
CelebAir (ITV 2) 163
Celebrity Big Brother
    (Channel 4) 232–9
Celebs Go Dating (E4) 324–5
Chalmers, Judith 167
Channel 4 170
    see also Big Breakfast, The;
    Celebrity Big Brother
Chapman, Nicky 324
Charles, Craig 318
Charles, Prince 126, 178
Charles, Tina 281
Chaucer, Geoffrey: *Troilus and Criseyde* 39
Chegwin, Keith 179
Chilcote, Judith 104
Clancy, Abbey 286

Clarence Hotel, Dublin 268
Clark, Dick 174
Clifford, Max 155
Clyne, Celia 98
CNN 146
Cohen, Ben 286
Cole, Michael 323
Cole, Tina 323
Comic Relief 232, 236, 284
Commitments, The (film) 328
Corbyn, Jeremy 260
Cosmopolitan (magazine) 77, 78
Costa, Antony 270
Covid 19 epidemic 292–7, 305
Cox, Sara 324
Crowne, Benjamin 84, 324
Crowne, Beverley 83, 84, 213, 324
Crowne, Joel 84
Crowne, Nicholas 324
Crowne, Samuel 84
Cukor, George 328
Currie, Edwina 155
Curtis, Richard 232

Daily Express 238, 248, 300
Daily Mail 125, 165,
    see also Mailonline
Daily Mirror 128, 131, 155, 200, 202, 214, 301–2
Daniels, Paul 242
dating 325–7
Davidson, Lorraine 43
Davie, Tim 318
Dawes, Marjorie 180

Day, Darren 176
Day, Elizabeth 324
Dayan, Moshe 269
Dee, Jack 234, 235, 236, 237
de Metz, Niki 323
Denmark Today 146–7
Depp, Johnny 100
Desmond, Richard 178
DeVito, Danny 142
Diana, Princess 35, 46, 163
DiCaprio, Leonardo 99–100, 107
Dolce, Joe 281
Donahue, Phil 106, 109, 110, 111
Doyle, Roddy 328
Duffy, Keith 234, 236, 237
Dury, Ian 31

Earle, Liz 324
Early Breakfast Show, The (Radio 2) 281, 282, 287, 296
Edinburgh, Duke of 89
Edmonds, Noel 176–7
Eilat, Israel: King Solomon's Palace 58–60
Eliot, T. S. 329
Elizabeth II 89, 90
   Golden Jubilee 320
   Platinum Jubilee 319–20
Ellis-Bextor, Sophie 286
Endemol UK Productions 238
Eubank, Chris 234, 235, 236–7
Eugenie, Princess 77
Evans, Chris 163, 275, 282
Evening Standard 39, 261
Everyone Says I Love You (film) 328
Evita (film) 146, 148
Eyjafjallajökull, Iceland 276

Fawlty Towers (Apollo Theatre) 332
Feltz, Babs (Grandma Babs) 10, 12, 19–20, 28, 57, 71, 84, 92–3
Feltz, Maurice (Mo) 18, 19, 20
Feltz, Norman 87, 268
   as golfer 7
   the 'Knicker King' 8, 125
   and Middle East crisis 26
   and Lord Jonathan Sacks 126–7
   and Valerie Feltz 20, 83–4, 124
   after Valerie's death 133, 181–2
   and Vanessa 9, 12, 15, 24, 42–3, 60, 64, 71, 72, 75, 85–6, 125, 127, 220, 229, 245, 252, 275–6
   and Vanessa's wedding 49, 51, 53, 54, 55, 57
Feltz, Valerie 8, 13, 35, 37–8, 80, 90, 191
   illness and death 123–4, 128–30, 131, 133, 222
   and Norman Feltz 20, 83–4, 124
   and Lord Jonathan Sacks 126–7
   and Vanessa 9–10, 11–16, 17, 32, 36–7, 62, 64, 71, 75, 90, 117, 125, 252
   and Vanessa's wedding 49, 50, 51, 52, 54–5, 56, 57, 58
Ferrari, Nick 320–21
Fiddler on the Roof (film) 19, 21, 219
Filippello, Connie 47

INDEX

Finnigan, Judy 79, 80, 81, 82, 84, 99, 131, 170, 183
Forsyth, Sir Bruce 286–7, 288
Fosse, Bob 328
Freedman, Freda 56
Freedman, Mick 56–7
French, Dawn 131–2, 162
Fry, Stephen 39, 40, 193
Fugard, Athol 39
Funny Girl (film) 100
Furnish, David 164

Gardner, Sara 167, 231
Gascoigne, Paul 178
gastric bands 277, 278–80, 290–91
gastric bypass 291–2, 307
Gaunt, Jon 252
Geldof, Bob 135, 179
Geldof, Fifi 179
Geldof, Peaches 179
Geldof, Pixie 179
George, Max 173
Gest, David 102, 163, 178
Gibb, Maurice 29
Gillard Award for Best Breakfast show 300
Girls Aloud 178
Gladstone, William 251
Global TV 320–21
Goddard, Trisha 197, 199
Gogglebox (Channel 4) 121
Gold, Anne-Marie 16
Goldblum, Jeff 143–4
golf clubs 7
Gooding, Cuba, Jr 143

Goodyear, Julie 106
Gove, Michael 253
Grant, Brigit 74
Grayson, Larry 69
Green, Charles 54
Greenspan, Mr (butcher) 95
Grenfell Tower fire (2017) 290
Guardian newspaper 118, 147, 153

Haberdashers' Girls' School 15–16, 27, 29, 33, 71
Hackenbrough, Miss (teacher) 72
Hair and Beauty (magazine) 70
Hair Flair (magazine) 70
Hammond, Alison 324
Harding, Sarah 178
Harrelson, Woody 139, 142
Harri, Guto 261
Harris, Alwen 150, 151, 152
Harris, Rolf 150–54
Harry, Prince 46
Harry's Bar, London 164
Hartwill, Andy 323
Hawn, Goldie 136, 140–42
Heaton, Michelle 178
Heineken advertisements 242–4
Hello! Magazine 213
Helmsley, Harry 118
Helmsley, Leona 118
Heston, Charlton 136
Hill, Jimmy 242
Hofman, Deborah 130
Hofman, Martin 95
houses, Vanessa's 57, 62, 63, 64, 117, 118, 183, 185–7, 188–9, 247, 329–30

Huberman, Amanda 130
Hudson, Kate 141
Hurley, Liz 164
Hutchence, Michael 134

If I Can, You Can (video) 229–30
Ifans, Rhys 180
Iglesias, Julio 281
Independence Day (film) 143
Inman, John 244
Instagram 81, 307, 315, 331, 333
Irons, Jeremy 39
Israel, Simon 95
It's a Wonderful Life (film) 248
ITV 80, 106, 192–3, 194, 195, 197
  see also *This Morning*; *Vanessa*;
  Vanessa's Real Lives

Jacklin, Tony 287
Jacobs, David 317
Jewish Burial Society 219
Jewish Chronicle 70–71, 74, 92,
  93, 97, 98, 213
Jewish London (GLR) 74, 75,
  127, 250
Jewish Telegraph 98
Jews 6–7, 8, 21, 55
Joel, Billy 207
John, Sir Elton 162
Johnson, Boris 260–61, 293,
  302, 320
Johnson, Stanley 293–4
Jonah and the Wailers 97
Jordan, James 285, 286, 287, 288
Jordan, Ola 286, 288

Joss, Amiel Jeffrey (AJ) 308–10, 313,
  324, 332–3
Joss, Cecily Violet 5, 311, 314, 332, 333
Joss, Marc 307, 311
Joss, Saskia Clemency 269, 270, 329
  and Allegra 64–5, 191
  batmitzvah 246, 247
  birth 64
  childhood 67, 72, 89, 118, 124, 131,
    147, 161, 167, 189, 205, 226,
    226–7, 245, 254
  and 'One Hit Wonder' 311, 314
  and parental divorce 205–6,
    208–9, 211, 212, 214, 215, 219
  school 212
  and Vanessa and Platinum Jubilee
    celebrations 319–20
  and Vanessa's appearance on
    Celebrity Big Brother 238, 239
  at Vanessa's 60th birthday
    celebrations 307
Jubilant (Royal barge) 320
Jurassic Park (film) 143

Kabs, Mr (from Flawless) 329
Kalman, Matthew 96, 97, 98
Kaplinsky, Natasha 285
Katona, Kerry 178
Katz, Alex 323
KC and the Sunshine Band 143
Kelsey, Linda 78
Kennedy, Sarah 281
Khan, Sadiq 261, 321
Kilroy-Silk, Robert: *Kilroy* 197
Klass, Myleene 324, 329

Knobil, Marcel 323
Kool & The Gang 143
Kovalev, Pasha 286

Lachter, Danny 323
Lamé, Amy 163
Lampon, Jonathan 262
Lancaster, Penny, Lady Stewart 162
Landau, Carole 130
Langdon, Ian 323
Langdon, Laura 323
Laurence, Barbara 89
Laurence, Captain Tim 89
Laurie, Hugh 39, 40
Lawson, Nigella 165
Layton, Pat 52
Lazarus, Nicholas 134, 137
Lazarus, Sheldon 137
Lederer, Helen: *Not That I'm Bitter* 278–9
Lemonia, Primrose Hill, London (restaurant) 163
Let's Dance for Comic Relief (BBC TV) 284
Liberty X 178
Liddiment, David 193
Lionel (hairdresser) 95
Lipman, Dame Maureen 159–60
Little, Mark 136, 148, 149, 160–61
Livingstone, Ken 257, 259–60
Lloyd Webber, Andrew 146
Loose Women (ITV) 268
Lucas, Matt 180–81
Lulu 29, 69
Lumley, Dame Joanna 262
Lush, Jane 193, 197
Lyons, Kate 275

McCutcheon, Martine 178, 264
Macdonald, Julien 287
McDonald, Sir Trevor 126
MacDowell, Andie 136
McGee, Debbie 242
Machado, Mark 323–4
MacLeod, Tracey 134
Madame Tussaud's, London 179
Madeley, Richard 79, 80, 81, 82, 84, 99, 103, 111, 131, 170, 183, 248
Madonna 145, 146–8
Maffia, Lisa 163
Maguire, Lorraine 301
Mailonline 165, 301, 312
Majesty magazine 46
Malka, Jonathan 18
Manchester Arena bombing (2017) 289
Manero, David 296
Mann, John 260
Marenbon, Professor John 39
Marine, Jeanne 179
Markle, Meghan 181
Marshall, Sharon 324
Mathis, Johnny 136
May, Theresa 290
Mayo, Simon 317–18
Meadows, Shane 180
Men Only (magazine) 70
Michael, George 47
Minnelli, Liza 178
Minogue, Kylie 254

Mirvis, Chief Rabbi Ephraim 270
'Miss Piggy' 148–9
Monroe, Marilyn 7
Morecambe (Eric) and Wise (Ernie) 170
Morgan, Piers 202, 214
Morrison, Anne 200
Mortimer, Bob 180
Moss, Peter 87, 88, 92, 95
Mother and Baby (magazine) 43
Motivator, Mr 178
Mott the Hoople 281
Muppet Treasure Island (film) 179
Murdoch, Rupert 319, 331
Murphy, Stuart 179

Nathan, Paul 27
National Magazines 77
New End Theatre, Hampstead, London 92–4, 96, 97
New Moon (magazine) 96
Newman, Harold 83
Newman, Marilyn 83
Nighy, Bill 163
Noakes, John 150
Nolan Sisters 281
Norrice Lea synagogue, London 219
Norton, Graham 136, 151, 170, 318

O'Grady, Paul 318–19
Ohrenstein, Sybil (Grandma Sybil) 6, 10, 12, 13–14, 19, 20, 44–5, 48–9, 50, 218, 323
Ohrenstein, William (Poppa Willie) 13, 20, 48, 75, 84, 218

OHW *see* 'One Hit Wonder'
OK! magazine 178, 264, 268, 278
'Old School Pal' 223–7
Once Upon a Time in the Midlands (film) 180
'One Hit Wonder' 263–73, 275, 279–80, 283, 294–5, 297–9, 300, 301–5, 306–9, 310, 311–16
One Leicester Square (MTV) 254
Orford, Richard 137
Orton, Clive 26, 27, 95, 231, 277
Orton, Melvyn 52
Oz, Frank 149

Page, Tony 98
Palin, Michael 332
Panto, Clive 97
Paris Match 146
Parker, Alan 146, 148, 328
Parker, Dorothy: 'One Perfect Rose' 329
Parkinson, Michael 139
Parsons, Charlie 137, 174
Patterson, Dan 97
Peel, John 282
Perry, Katy 254
Peter's Friends (film) 40
Philadelphia Story, The (film) 328
Phillips, Fiona 323
Phillips, Gary 323
Pinter, Harold 267
Piper, Billie 163
Planet 24 (company) 134, 135, 137, 152
Plant, Linda 331
Plantin, Marcus 106, 112, 115–16, 118

INDEX

Plaskow, Rabbi Michael 18
Pollard, Eve 241
Poole, Professor Adrian 35–6
Porter, Cole 329
Portsmouth: King's Theatre 308–10
Poseidon Adventure, The (film) 290
Povich, Maury 102–3, 109, 110, 169
Price, Katie 178, 179
'Prince Charmless' 29–34, 40–41, 51

Quaid, Dennis 139

Rabbitte, Jimmy 328
Raffa-Rickman, Cynthia 110–11, 132–3, 135, 208
Ramsay, Gordon 252
Rantzen, Dame Esther 246
Raphael, Sally Jessy 106, 109, 110
Raymond, Paul 70
Redwood, John 180
Reeves, Vic 180
Reid, Susanna 286
Reuben, Linda 323
Reuben, Mark 298, 323
Rice, Anneka 288
Rice, Tim 146, 329
Richard, Cliff 69
Richie, Shane 248
Ridgeley, Andrew 47
Rihanoff, Kristina 286
Riley, Lisa 242, 244
Riley, Rachel 286
Rinder, Rob 288
Rivers, Joan 136, 140
Robson, Linda 324

Rooney, Coleen 223
Ross, Jonathan 151, 212, 254
Ryan, Meg 139
Rylan 324
Ryvita advertisements 277

Saatchi, Charles 165
Sachs, Andrew 254
Sacks, Rabbi Lord Jonathan 126–7
Salmon, Peter 193, 197, 198, 202
Saunders, Jennifer 179
Savage, Lily 134, 135
Savile, Sir Jimmy 242–3
Sayer, Leo 281
Scherzinger, Nicole 285
Schneider, Dave 89, 97
Schofield, Phillip 282, 324
Sedlackova, Monika 91
Segen, Michael 53
7/7 bombing (2005) 257–9
Sexton, David 39
Shah, Naz 260
Shalit, Katrina 331
Sharif, Omar 100
She (magazine) 77, 78, 102
Shear, Dalya 323
Shear, Graham 175, 194, 195, 196
Shelley, Andrew 38
Shine, Rabbi 55, 56
Shmuley, Cousin 270
Shure, Jan 213
Silverstone, Alicia 136
Singleton, Valerie 150
Škorjanec, Aljaž 286
Snipes, Wesley 142–3

Somers, Mr Shaw 291
Sony Gold Awards: Speech Personality of the Year (2009) 275–6
Spake, Jeremy 179
Spice Girls, the 176
Springer, Jerry 111, 120, 169, 170, 171–2
Springfield, Dusty 69
Stallone, Sylvester 100
Stanhope, Esther 257
Stapleton, John 100–1, 103, 106–7
Staunton, Imelda 40
Steinberg, Gillian 86
Stevenson, Gemma 301
Stewart, Sir Rod 162
Stoller, Sam 80, 92, 97
Streisand, Barbra 100
Strictly Come Dancing (BBC TV) 285–8
Stringfellow, Peter 242, 243–4
Sugar, Lord Alan 331
Sun newspaper 78, 118, 325
Sunak, Rishi 331
Sunday Express 153
Sunday Love Songs (BBC Radio 2) 318
Survivor (ITV) 137
Sweeney, Claire 234, 236, 237, 324
Swinton, Tilda 39

Take That 298
Tarrant, Chris 212
Temko, Ned 70

This Morning (ITV) 79, 80–82, 84, 98, 99–100, 104, 131, 183, 277, 282, 306–7
Thomas, Dylan: 'Under Milkwood' 329
Thomas, Helen 296, 317
Thompson, Emma 39, 40
Thorntons chocolate shops 187
Time, The Place, The 100–1
Times, The 118, 253
Times of Canada 146
Tomlinson, Ricky 180
Tong, Peter 23–4
Top of the Pops (BBC TV) 263, 264
Totteridge, North London 6–7
Trinity College, Cambridge 33, 35–6, 38–9, 42, 219
Tuke, Simon 38
Turner, Anthea 234, 235–7, 288

University College Hospital, London 63

Value for Money (BBC TV) 128, 132, 155, 192
Van Outen, Denise 324
Vandross, Luther 281
Vanessa (ITV) 106, 107–8, 112, 113, 114–16, 118, 120–22, 125, 128, 131, 132, 155, 183, 192, 194, 196, 197
Vanessa Show, The (BBC TV) 193–5, 196–7, 197–9, 200–2, 210
Vanessa's Day With . . . ? (Channel 5) 155, 160, 192

Vanessa's Real Lives (ITV) 266
Vaughan, Frankie 7–8, 69
Victoria, Queen 251
Vine, Jeremy 281
Vorderman, Carol 193

Walliams, David 180–81
Wanted, The 163
Watson, Zoe 109, 111
Waugh, Evelyn: *Brideshead Revisited* 39
Weinstein, Harvey 154
Weller, Paul 83
Wengrowe, Dr Nolan 95
Wham! 47
What's Eating Gilbert Grape? 99
When Harry Met Sally (film) 139
Whiley, Jo 318
Widdecombe, Ann 285
Wilkin, Andrea 52, 56
William, Prince: wedding 320
Williams, Kenneth 55
Williams, Robbie 298
Willoughby, Holly 282, 324
Wilmot, Gary 244
Winfrey, Oprah 106, 107, 120, 241

Winkleman, Claudia 241
Winner, Michael 247
Winters, Shelley 290
Winton, Dale 208
Wish You Were Here . . . ? (ITV) 167–8
Wogan, Sir Terry 281, 328
Woking, Surrey: New Victoria Theatre 245
Woman's Own 118
Wonder, Stevie 281
Wright, Amelia 283
Wright, Matthew 166–7, 275, 283
Wright, Steve 318

Yates, Paula 134, 135
Yom Kippur 24–5
York, Sarah Ferguson, Duchess of 77
You Bet (BBC TV) 176
Young, Sir Jimmy 317
Young, Paul 155

Zavaroni, Lena 69
Zephaniah, Benjamin 329
Zeta-Jones, Catherine 134